Invasive Species - Introduction Pathways, Economic Impact, and Possible Management Options

Edited by Hamadttu El-Shafie

Published in London, United Kingdom

IntechOpen

Supporting open minds since 2005

Invasive Species - Introduction Pathways, Economic Impact, and Possible Management Options
http://dx.doi.org/10.5772/intechopen.87502
Edited by Hamadttu El-Shafie

Scientific Contributors
Marcos Soto-Hernández, Jorge Reyes-Rivera, Agustin Maceda-Rodríguez, Mariana Palma-Tenango

Contributors
Maria Galkina, Yulia Vinogradova, Francisca Hernández, Lucia Andreu Coll, Marina Cano-Lamadrid, David Lopez Lluch, Ángel Carbonell Barrachina, Pilar Legua Murcia, Swoyam Singh, Hashib Ansari, Jai Hind Sharma, Avinash Udiker, Younes Moussaoui, Faten Mannai, Ramzi Khiari, Prof. Dr. Muzammil H. Siddiqui, Karina Buzzetti, Hamadttu El-Shafie, Maged E. A. Mohammed, Mohammed Alhajhoj, Jose Romeno Faleiro, Muhammad Aamir Iqbal, Sajid Ali, Ayman El Sabagh, Zahoor Ahmad

First published in London, United Kingdom, 2020 by IntechOpen
IntechOpen is the global imprint of INTECHOPEN LIMITED, registered in England and Wales, registration number: 11086078, 5 Princes Gate Court, London, SW7 2QJ, United Kingdom
Printed in Croatia

British Library Cataloguing-in-Publication Data
A catalogue record for this book is available from the British Library

Additional hard and PDF copies can be obtained from orders@intechopen.com

Invasive Species - Introduction Pathways, Economic Impact, and Possible Management Options
Edited by Hamadttu El-Shafie
p. cm.
Print ISBN 978-1-78985-849-5
Online ISBN 978-1-78985-850-1
eBook (PDF) ISBN 978-1-83880-431-2

For EU product safety concerns:
IN TECH d.o.o., Prolaz Marije Krucifikse Kozulić 3, 51000 Rijeka, Croatia, info@intechopen.com or visit our website at intechopen.com.

Meet the editor

Hamadttu Abdel Farag El-Shafie is Associate Professor of Entomology and Senior Research Entomologist at the Date Palm Research Center of Excellence, King Faisal University, Saudi Arabia. He is also head of the IPM research program at the same researcher center. Dr. El-Shafie obtained his BSc and MSc from the University of Khartoum Sudan in 1988 and 1993, respectively. He received his PhD from the University of Giessen, Germany, in 2001. In 2008, he was appointed head of the Crop Protection Department, and then deputy dean for academic affairs at the Faculty of Agriculture, University of Khartoum. He supervised twenty-five MSc students and five PhD students at the University of Khartoum. His research interests focus on management of field crop pests using neem biopesticides and semiochemicals. Dr. El-Shafie has more than ten years' experience in management of the invasive red palm weevil and other date palm pests. He has published sixty research papers in international peer-reviewed journals and fourteen book chapters with international publishers such as Springer, John Wiley & Sons, and IntechOpen, in addition to more than twenty-five international conferences in the field of entomology. During the last ten years, he has been reviewing manuscripts for thirty renowned international journals.

Contents

Preface **XI**

Section 1
Invasive Insect Species **1**

Chapter 1 **3**
Red Palm Weevil *Rhynchophorus ferrugineus* (Coleoptera: Curculionidae): Global Invasion, Current Management Options, Challenges and Future Prospects
by Hamadttu Abdel Farag El-Shafie and Jose Romeno Faleiro

Chapter 2 **33**
Recent Trends in the Early Detection of the Invasive Red Palm Weevil, *Rhynchophorus ferrugineus* (Olivier)
by Maged E.A. Mohammed, Hamadttu A.F. El-Shafie and Mohammed R. Alhajhoj

Chapter 3 **49**
Invasive Insects in India
by Swoyam Singh, Jai Hind Sharma, Avinash Udikeri and Hashib Ansari

Chapter 4 **65**
The Spotted Wing Drosophila in the South of the World: Chilean Case and Its First Productive Impacts
by Karina Buzzetti Morales

Chapter 5 **77**
Tuta absoluta (Meyrick) (Lepidoptera: Gelechiidae): An Invasive Insect Pest Threatening the World Tomato Production
by Hamadttu Abdel Farag El-Shafie

Section 2
Invasive Plant Species **93**

Chapter 6 **95**
Hybridogenic Activity of Invasive Species of Asteraceae
by Maria A. Galkina and Yulia K. Vinogradova

Chapter 7 113
Changing Climate and Advances on Weeds Utilization as Forage: Provisions, Nutritional Quality and Implications
by Muhammad Aamir Iqbal, Sajid Ali, Ayman El Sabagh, Zahoor Ahmad and Muzammil H. Siddiqui

Chapter 8 127
Valorization of Prickly Pear [*Opuntia ficus-indica* (L.) Mill]: Nutritional Composition, Functional Properties and Economic Aspects
by Francisca Hernández García, Lucia Andreu Coll, Marina Cano-Lamadrid, David López Lluch, Ángel A. Carbonell Barrachina and Pilar Legua Murcia

Chapter 9 143
Novel Trend in the Use of *Opuntia* (Cactaceae) Fibers as Potential Feedstock for Material Science Applications
by Faten Mannai, Ramzi Khiari and Younes Moussaoui

Preface

Invasive alien species (IAS) are non-native species to a particular locality and have the tendency to spread very quickly when they have successfully established themselves in the new invaded areas. IAS include plants, insects, birds, nematodes, fish, frogs, cats, snakes, turtles, toads, and pathogens. The stages of invasiveness of these species include introduction pathways, establishment, and dispersal. Invasive species are found in all ecosystems including terrestrial, marine, and other aquatic ecosystems, and they can be introduced into new areas either accidentally or intentionally through anthropogenic activities. Introduction pathways of IAS include agricultural trade and accidental transport of adults and pre-imaginal stages of insects, plant seeds, and vegetative parts in containers and vehicles. Infested fruits and planting materials are the key introduction pathways for several invasive insect species including palm weevils, fruit flies, emerald ash borer (EAB), mealy bugs, spider mites, western flower thrips, tomato pinworm, the brown marmorated stinkbug, the fall armyworm, and others. IAS can spread by cars, trucks, ships, aircrafts, and other means of travel and transportation of goods. Increasing trade through globalization facilitates the spread of invasive species. Climate change also hastens the spread of IAS and renders ecosystems more vulnerable to invasion by non-native or exotic species. Insects are among the most numerous invasive species worldwide and the number of new invasions is increasing exponentially, representing a threat to the economy and the environment. Insects represent about 87% of invasive species and cost the global economy more than 70 billion US$ annually, according to the International Union for Conservation of Nature (IUCN). It is extremely difficult and sometimes near impossible to eradicate invasive species completely once they have established themselves in a new habitat. In most cases, exotic species flourish in their new habitat due to prevailing congenital environmental conditions and the absence of their natural enemies.

The impact of IAS on ecosystems may include endangering biological diversity through completion and niche displacement, vectoring of serious human and animal diseases, and increasing the risk of native species extinction. Inter-regional immigration through introduction of alien plant species may lead to hybridization, which is likely to increase the threat of species extinction due to introgression. IAS could also alter phylogenetic diversity across communities, modify trophic chains, and change ecosystem functioning and delivery of ecosystem services by changing nutrients and contaminants' cycling.

In the time of globalization, invasive species are gaining an increasing economic and ecological significance. There is a need for more research and comprehensive edited volumes on the impact of IAS on global agricultural activities as well as on the different ecosystems, particularly with the increasing evidence of climate change and global warming. This book addresses issues pertaining to invasive insect and plant species with emphasis on their introduction pathways, bionomics, dispersal, risk assessment, economic impact, possible management, and control options.

The book contains nine chapters over two sections; section one is about invasive insect species and includes five chapters, while section two deals with invasive plant species and contains four chapters. Sections one gives comprehensive information on the red palm weevil, which has emerged as the most serious and damaging pest of palm species grown for production of dates, coconuts, and oil. The Canary Island palm, mainly used for landscaping, is also severely injured by palm weevils. Included in this section are the tomato pinworm, which is threatening tomato production worldwide, and the emerging spotted wing *Drosophila*. Section two contains four chapters dealing with the hybridogenic activity of invasive species of Asteraceae and economic utilization of invasive weeds as forages and animal feed. The section also contain two chapters dealing with the economic aspects of some invasive species in the family Cactaceae, namely, *Opuntia ficus-indica*. Economic utilization of invasive species represents one important option for their management and raises the debatable question of whether invasive species are friends or foes. The information included in this book will benefit researchers, university teachers, ecologists, students, policymakers, and environmental activists.

I would like to thank all chapter authors for writing and submitting their excellent works. It would have been difficult to produce this book without their contributions. Other contributors who deserve thanks and acknowledgements are Author Service Manger Dolores Kuzelj, Commissioning Editor Kristina Jurdana, and other staff at IntechOpen for their unlimited assistance during the preparation of this book. I would also like to thank Marcos Soto-Hernandez, Jorge Reyes-Rivera, Agustin Maceda-Rodriguez, and Mariana Palma-Tenango who helped by reviewing and editing Chapters 8 and 9. The effort exerted by Commissioning Editor Martina Usljebrka Kauric at IntechOpen in reviewing Chapters 8 and 9 is appreciated and sincerely acknowledged.

Sincere and heartfelt thanks are due to my wife Nawal, son Ayman, and daughters Hiba, Hala, Safa, and Lojain for their understanding, continuous support, and encouragement. I dedicate this book to my colleagues at the Date Palm Research Center of Excellence, King Faisal University, and my colleagues, students, and friends at the Department of Crop Protection, University of Khartoum, Sudan.

Dr. Hamadttu A. F. El-Shafie
Research Entomologist,
Date Palm Research Center of Excellence,
King Faisal University,
Al-Ahsa, Saudi Arabia

Department of Crop Protection,
University of Khartoum,
Shambat, Sudan

Section 1

Invasive Insect Species

Chapter 1

Red Palm Weevil *Rhynchophorus ferrugineus* (Coleoptera: Curculionidae): Global Invasion, Current Management Options, Challenges and Future Prospects

Hamadttu Abdel Farag El-Shafie and Jose Romeno Faleiro

Abstract

The red palm weevil (RPW) *Rhynchophorus ferrugineus* (Coleoptera: Curculionidae) also known as the Asian palm weevil is a key pest of palms (Arecaceae) in diverse agro-ecosystems the world over. During March 2017, the Food and Agricultural Organization of the UN through its 'Rome Declaration' called for the urgent need to combat RPW by collaborative efforts and commitments at the country, regional and global levels to stop the spread of this devastating pest. There exist gaps and challenges in almost all the components of the current RPW-IPM strategy, particularly with regard to early detection, developing and implementing phytosanitary measures, lack of effective biological control agents in the field and poor farmer participation in the control programmes, which have made RPW control and eradication extremely difficult. This chapter gives an overview of the global invasion, current management options, challenges and future prospects for its effective control.

Keywords: invasive species, red palm weevil, IPM, introduction pathways, quarantine insect

1. Introduction

The red palm weevil (RPW) *Rhynchophorus ferrugineus* (Coleoptera: Curculionidae) also known as the Asian palm weevil is an invasive key pest of palms (Arecaceae) in diverse agro-ecosystems the world over. The Food and Agriculture Organization of the UN has designated RPW as a category-1 pest in the Middle East and North Africa (NENA region), where it is a threat to the livelihood security of date palm farmers in rural communities [1]. During March 2017, FAO organized a 'Scientific and High-Level Meeting on the Management of RPW' and through 'Rome Declaration' called for the urgent need to combat RPW by collaborative efforts and commitments at the country, regional and global levels to stop the spread of this devastating pest. The pest has its home in South and South East Asia, where it is a key pest of coconut, *Cocos nucifera*. The cryptic behaviour and the intrinsic biological traits of the weevil have made it difficult to detect and therefore difficult to manage. Now, there are so many gaps and challenges in the components of RPW management strategies. These include early detection of the weevil infestation, limitation of its biological control agents under field conditions and lack of farmers' participation in the control operations [2]. The aim of this chapter is to give a consolidated information on gaps and challenges in current RPW management as well as future prospects.

2. Global RPW invasion: an update

During the mid-1980s, RPW was reported on date palm *Phoenix dactylifera* L. from the United Arab Emirates in the Gulf region of the Middle East. Subsequently, first reports of RPW invasion came from the other Gulf countries of the Middle East. During 1993, RPW attack was reported on date palm in Egypt in North Africa, and later during 1995, it was detected for the first time on *P. canariensis* from Spain in Europe. During the next two decades, the pest spread rapidly in the Gulf region of the Middle East, some Maghreb countries in North Africa and the Mediterranean basin countries in Europe (**Figure 1**). RPW has rapidly expanded its geographical range during the last three decades, and ecological niche modeling [3] suggests that the pest is likely to expand its geographical range still further. Recent reports of RPW invasion suggest that the pest is establishing in East Africa in Djibouti on date palm and also in the Caucasian region where it is detected in Abkhazia on the canary island palm in the Republic of Georgia. During 2019, RPW was detected in Bulgaria in the Black Sea Basin region and also in Bosnia-Herzegovina in Southeastern Europe.

The only report of RPW invasion in the American continent comes from Aruba and Curacao islands in the Caribbean region [4]. Although RPW was reported from California, USA, during 2010 [5], molecular studies at the University of California, USA, subsequently characterized the pest as *R. vulneratus* [6], a closely related species of *R. ferrugineus* predominant on coconut in the South East Asia. Similarly, the previous *R. ferrugineus* reports from Australia and countries in the Oceanic region have now been attributed to other species of the *Rhynchophorus* group of weevils. It is pertinent to point out that although RPW moves through infested offshoots in the date palm-growing countries, another important route of transmission/movement of the pest is through palms shipped for ornamental gardening.

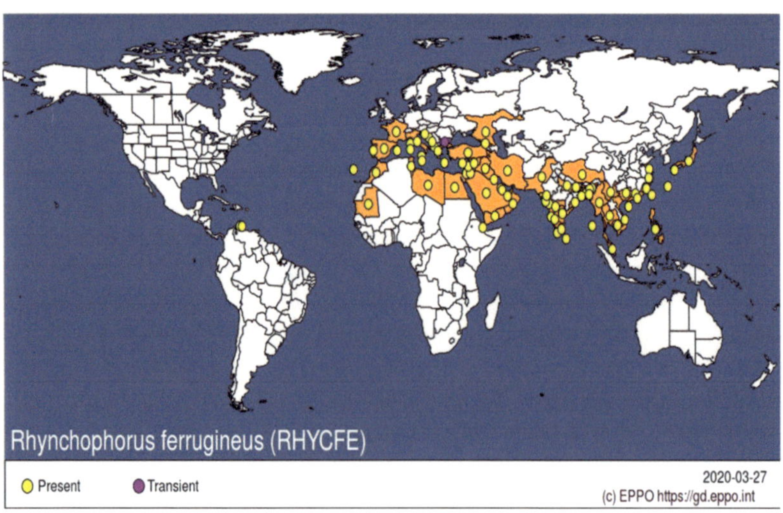

Figure 1.
Current Geographical Distribution of RPW (Source- EPPO 2020; https://gd.eppo.int/taxon/RHYCFE/datasheet).

The recent *R. ferrugineus* data sheets of the European Plant Protection Organization [7] reports that RPW exists in 49 countries including 15 in Europe (Albania, Bosnia-Herzegovina, Croatia, Cyprus, France, Georgia, Greece, Italy, Malta, Montenegro, Morocco, Portugal, Russia, Spain and Turkey), six in Africa (Djibouti, Egypt, Libya, Mauritania, Morocco and Tunisia), 26 in Asia (Bahrain, Bangladesh, Cambodia, China, India, Iran, Iraq, Israel, Japan, Jordan, Kuwait, Lebanon, Malaysia, Myanmar, Oman, Pakistan, the Philippines, Qatar, Saudi Arabia, Sri Lanka, Syria, Taiwan, Thailand, the United Arab Emirates, Vietnam and Yemen) and two in the American continent (Aruba and the Netherlands Antilles).

3. Host range and introduction pathways

As the geographical range of the pest expanded, so did the host range of RPW. During the mid-1950s, RPW was reported from just four palm species, viz. *C. nucifera*, *P. dactylifera, Metroxylon sagu* and *Corypha umbraculifera* [8]. RPW is currently reported on 40 palm species worldwide [7, 9, 10], including *Areca catechu* L., *Arenga saccharifera* Labill, *A. engleri* Becc., *A. pinnata* (Wurmb), *Bismarckia nobilis* Hildebrand and Wend, *Borassus flabellifer* L., *B.* sp., *Brahea armata S. Watson*, *B. edulis*, *Butia capitata* (Mart.) Becc., *Calamus merrillii* Becc., *Caryota cumingii* Lodd., *C. maxima* Blume, *Cocos nucifera*, *Corypha utan* Lamk., (= *C. gebanga*, *C. elata*), *C. umbraculifera* L., *Chamæerops humilis*, *Elaeis guineensis*, *Livistona australis* (R.Br.) Mart., *L. decipiens* Becc., *L. chinensis* Jacq.R. Br., *L. saribus* (= *L. cochinchinensis*) (Lour.) Merr., *Metroxylon sagu* Rottb., *Oncosperma horrida* (Scheff.), *O. tigillarium* (Ridl.), *Phoenix canariensis* (Chabaud), *P. dactylifera*, *P. roebelinii* O'Brien, *P. sylvestris* Roxb, *P. theophrastii* Greuter, *Pritchardia pacifica* Seemann and Wendland, *P. hillebrandii* (Kuntze) Becc., *Ravenea rivularis* Jumelle and Perrier, *Roystonea regia* (Kunth.), *Sabal umbraculifera* (Jacq.) Martius, *Trachycarpus fortunei* (Hook), *Washingtonia filifera* (L. Lindl), *W. robusta* H. Wendl. and *Syagrus romanzoffiana* (Cham.). The non-palm hosts are the century plant *Agave americana* and sugarcane *Saccharum officinarum* [9].

Considering such a wide host range of this pest, it is imperative to ensure proper phytosanitary measures are in place before palms are transported/shipped. Recently, FAO has published the detailed guidelines on phytosanitary regulations to be adopted to ensure the movement of RPW free planting material within national borders and from one country to another [11]. Import and movement of infested plant material within a country are the main pathways to the introduction and spread of *R. ferrugineus*. Short-distance spread is possible by adult flight. The pest can be spread over long distances in infested palms for planting of host palms [7]. Flight mill studies have demonstrated that RPW being a sizeable population is short-distance fliers (<100 m) which would explain the aggregated/clumped distribution of infestation. RPW has the capacity to fly up to 50 km in a day with flight activity being predominantly diurnal [12–14].

4. Biology and symptoms of damage

Understanding the biology of the pest is essential in developing, implementing and sustaining management practices. RPW is a hidden pest, with all its life stages developing inside the palm except for the adult stage, which is partly exposed when adult weevils fly out of the brood in search of the host or on occasions to find a mate.

Recently, Al-Ayedh [15] in the FAO guidelines on RPW management and EPPO [7], using RPW data sheets, has summarized the literature on the biology of this pest.

There are several previous reports on the biology of RPW on natural and artificial diets [8, 16–23]. **Figure 2** depicts the life stages with the probable duration of each stage.

RPW takes about 3–4 months to complete its life cycle. Volatiles released from fresh wounds/cuts on the palm helps in egg laying by attracting adult RPW female weevils. Gravid females lay eggs in cracks and crevices on soft palm tissue. In coconut and date palm, oviposition usually occurs in young palms below 20 years old. There is a weak relationship between *Oryctes* sp. infestations and RPW attack in coconut and date palm [24, 25]. An adult female lays over 200 eggs. Oviposition is strongly affected by temperature [22]. On hatching, the legless larvae start feeding and move towards the interior of the palm. In areas with a mean annual temperature (MAT) below 15°C, one generation per year can be expected, while more than two generations in those with MAT above 19°C. Several overlapping generations of the pest may occur inside a single infested palm. Further, in the Mediterranean region, the larval stage can get prolonged up to 160 days in winter-spring seasons [22]. Upon completion of the larval period (7–16 instars), mature larvae pupate in cylindrical fibrous cocoons, leading to the emergence of adult weevils (**Figure 2**). When RPW was reared in the laboratory on a meridic diet, some of the larvae successfully pupate and develop to adults without construction of cocoons (**Figure 3**) (El-Shafie, unpublished data).

Date palms below the age of 20 are more susceptible to attach by RPW. Infestation is found more common on the trunk within 1 m from the ground; however, infestation can occurs on aerial offshoots and the crown of male palms [26]. Damage symptoms on Date palm (*Phoenix dactylifera*) have been described in detail [26–28]. The main symptoms include the following:

i. Oozing of brownish viscous fluid together with frass (palm tissue excreted by feeding grubs) which has a typical fermented odour

ii. Drying of outer leaves and fruit bunches and drying of infested offshoots

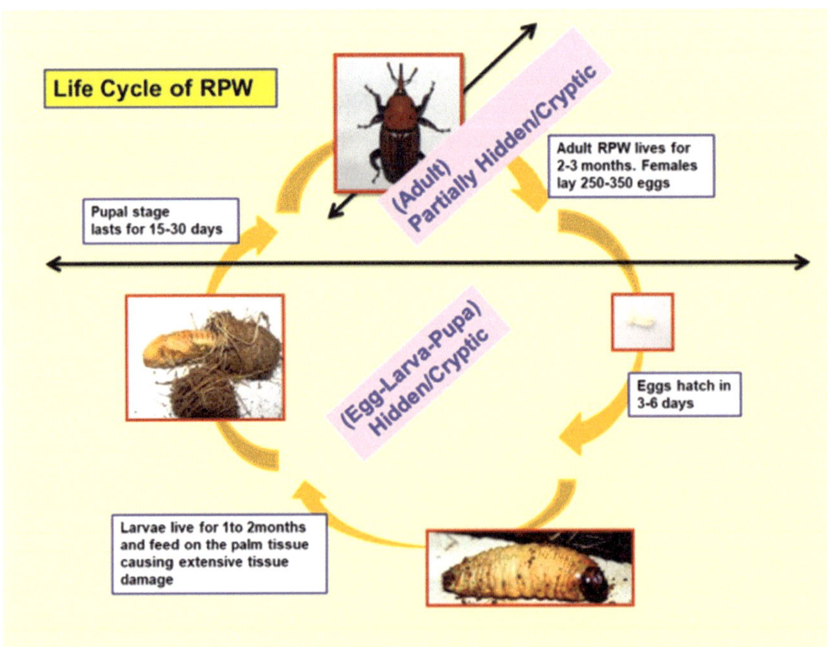

Figure 2.
Schematic Diagram Depicting the Life Cycle of RPW (Source: Faleiro and Al-Shawaf 2018; http://hdl.handle.net/20.500.11766/8914).

iii. Topping of the trunk in case of severe and extensive tissue damage

iv. Presence of adults and pupae at the base of the fronds and on the ground near infested palms

Figure 3.
Successful emergence of a viable RPW adult from a pupa without fibrous cocoon (Photo: Hamadttu A. F. El-Shafie).

Figure 4.
Drying or withering of aerial offshoots, the fronds of which can be easily pulled out (left). Oozing out of brown viscous unpleasant liquid from wounds in the trunk (right) (Photo: Hamadttu A. F. El-Shafie).

Figure 5.
Ejection of chewed-up fibers from the crown of an infested young date palm (left). Smashed internal palm tissue due to feeding of grubs (Photo: Hamadttu A. F. El-Shafie).

Visual symptoms of damage on palms are used for early detection of RPW infestation. Thus, it is of paramount importance for field inspectors and palm growers to understand these symptoms (**Figures 4** and **5**). On Canary Island palm (*Phoenix canariensis*), infestation and damage occur in the crown. The larvae tunnel in the developing leaves (fronds) and severe infestation may lead to palm mortality. Early symptoms include the presence of holes in the fronds, which look chewed and broken. Wilting and drying of developed fronds and absence of new emerging fronds cause asymmetrical growth of the crown that later collapse. As is the case with date palm, different stages of the weevil can be seen at the crown particularly when infestation is more severe [29].

5. Current RPW management programmes

In areas where the pest does not exist but under the potential risk of infestation occurring, it is essential to emphasize on quarantine, monitoring/surveillance and capacity building.

A new infestation report calls for immediate removal and destruction (eradication) of the RPW infested palm right at the sight/farm where the infestation is detected. Subsequently, a surveillance programme based on a regular inspection to detect infestation and monitor trapping to capture emerging adults needs in the demarcated area to ensure effective control, containment and eradication of the pest. Chouibani [11] in the FAO guidelines proposed to identify the infested zone where the presence of RPW is confirmed and also a buffer zone extending at least 10 km beyond the boundary of the infested zone. A strict vigil is to be maintained on the movement of palms and plant nurseries within the demarcated area. The demarcated area will be declared free from RPW if, during three consecutive years, RPW has not been detected [11].

Geographic information system (GIS) provides a very valuable tool in monitoring, predicting, managing and fighting the spread of pests and diseases, and

GIS-based techniques are increasingly used to enhance and support decision-making capabilities in RPW management [30–33]. This tool offers opportunities for cost-effective and efficient targeting of control interventions. In monitoring, GIS can be used to determine the spatial extent of a pest, to predict the projected spread, to provide input for risk assessment models. The first and essential step for efficient use of GIS techniques at a larger scale is a protocol for data collection. This will help to have a grip of the situation and is essential to periodically validate the RPW control programme, where data on infestation reports and trap captures are important. Fajardo et al. [33] proposed the following with regard to the use of GIS in area-wide RPW management programmes:

- Maintain a field map of each operational area indicating the basic data (number of palms, year of planting, number of traps, number of infested palms, palms removed, palms treated, etc.).

- Register the GPS co-ordinates of all the palms in a geographical data base if possible. The more important fields to be included are: date, area, height and state (not infested, infested and palms removed with GPS co-ordinates). It is also recommended to register the coordinates of the already removed palm trees.

- Data on the geo-reference localization of the palms, the RPW-IPM components and their evolution over time using GIS to elaborate maps and analysis need to be developed.

- FAO has proposed a real-time database and a web portal for the management of RPW at the local, national and Near East and North Africa (NENA) region. Furthermore, a mobile app for android and iOS smartphones to record geo-referenced data at the field location on a standard form needs to be developed. FAO has made initiatives in this regard both at the regional (NENA) and global levels [34, 35].

In any area-wide IPM programmes, the means (resources) to control the pest can be correlated with the intensity of the pest. Ferry et al. [36] visualized three scenarios to exist in the current RPW-IPM strategy depending on the resources available to control the RPW, considering that the organization and techniques are optimum and similar for the three scenarios:

1. The means are superior to the needs: here the resources are adequate and the pest is controlled/eradicated.

2. The means remain more or less equal to the needs: here there is a prolonged effort to control the pest over several years with little or no success and the pest is always ahead with the IPM strategy trying to catch up. Such a scenario is not sustainable in the long run.

3. The means are inferior to the needs: here the pest is not controlled and proliferates rapidly. The control means are inadequate.

It has been seen is several countries that providing adequate man power and material is a major challenge in all area-wide RPW control programmes. It is essential to provide adequate resources right at the beginning of the first record of this pest so that RPW can be efficiently controlled and eradicated when the pest is confined to a few farms/small area.

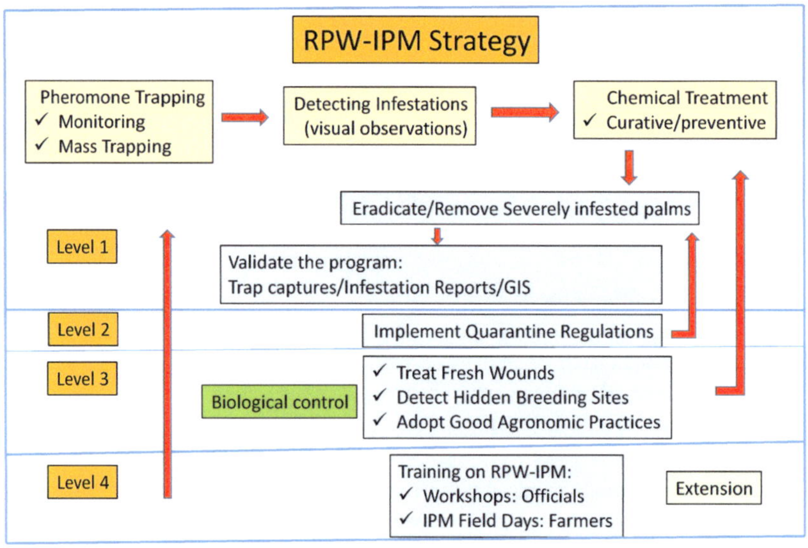

Figure 6.
RPW-IPM strategy implemented (Updated from http://www.fao.org/3/a-ms665e.pdf).

Figure 6 depicts the components of the current RPW-IPM strategy at four levels. At level-1, the strategy realizes the control components at the operational level on a daily basis in the field. The area-wide management of RPW needs careful planning and timely intervention of the control techniques, able supervision and periodic performance analysis of the RPW-IPM strategy, besides the desired technical, human, intuitional, organizational and coordination capacities for effective planning, delivery, monitoring and management of RPW in the field.

At *level-1* of the strategy, the RPW-IPM components connected with the day-to-day operations in the field are highlighted [37].

5.1 Detection of infested palms

Success of an RPW-IPM programme lies in the early detection of infested palms, and currently, visual inspection of palms is widely adopted to locate infested palms. Here, it is essential to break the cycle of the pest by locating an infested palm before adults emerge. A well-trained person can inspect 200–300 date palms per day depending on the terrain, palm density and field sanitation techniques adopted. In this context, a regular 45-day interval inspection of date palms in the susceptible age group of less than 20 years old is necessary. Vidyasagar [38] and Jaques [29] have detailed the protocols for visual inspection of the date and Canary island palms, respectively, in the FAO guidelines for RPW management. Research is underway in several countries to develop a cost-effective and user-friendly early detection device. Advanced techniques such as detecting chemical signatures, acoustic detection, use of infrared cameras, thermal imaging and satellite imaging/IoT are being researched upon [39–43]. However, farmers have to rely on visual (manual) inspection to detect an RPW infested palm, as these techniques are limited by their cost and the need for installing sophisticated hardware that is not easy to operate and requires specialized staff to operate.

5.2 Pheromone trapping

Ever since Hallett et al., [44] discovered the male-produced aggregation pheromone (ferrugineol) for RPW, food-baited (natural kairomone) bucket traps have been widely used in monitoring and mass trapping programmes in the field. Over the years, trapping protocols with respect to trap design, trap colour, lures, release rate, food bait, trap servicing (periodic change of food bait), role of co-attractants, etc. have been researched in several countries [29, 45–50]. Pheromone trap captures help optimizing/prioritizing the inspection of palms to detect infestations. Depending on the availability of human resources, palms around the traps with higher weevil captures should be inspected on priority. El-Shafie and Faleiro [51] reported through the controlled olfactometer studies that only a part of the adult population is attracted to the pheromone lure, which calls for the integration of pheromone trapping with other RPW-IPM techniques. There is a tendency of over-dependency on pheromone trapping and neglecting other RPW-IPM components, which leads to the build up and proliferation of the pest. Although the four-window black coloured bucket traps are popular, the dome-shaped conical Picusan™ is also used in several European countries.

It is of utmost importance to adopt the best trapping protocols with respect to trap design, trap colour, density, servicing (periodic renewal of food bait), trap placement, lure attraction and longevity, etc., for food-baited RPW pheromone traps. Sub-standard trapping protocols would adversely impact the trapping efficiency and consequently limit the success of the control programme [52, 53]. In some countries, the food bait and water are placed in a small container inside the bucket trap. Often this container is insufficient to hold the required amount of the bait or falls inside the bucket, emptying the water resulting in the food becoming dry, which consequently adversely impacts the performance of the trap due to poor bait-lure synergy. It is therefore recommended to place the required amount of the food bait (150–200 g of dates) directly in the water (1 L) inside the bucket trap. Of utmost importance is the fortnightly servicing (replacement of the food bait and water) in the trap. An economic rationale would demand that easily available locally sourced food baits with good attraction to be used as bait in the trap. Consequently, green coconut petiole pieces would do well in the coconut growing countries of South and South East Asia, low-grade dates in the Middle East and North Africa and palm tissue/petiole pieces in the Mediterranean region where the Canary Island palm is popular. Pheromone lures are known to last for 2–3 months in the field. Adding a small amount (1 g) of non-repellent insecticide granules (carbofuran/lanate) to the water in the trap could be useful in preventing the escape of adult weevils that enter the trap. However, in several countries where thousands of traps are in the field, the addition of insecticide in the trap is not practiced in view of the toxic side effects to the environment and potential harm to the staff who service the traps. Pheromone lures with both high attraction and field longevity should be selected. Do not discard old lures in the field or carry old lures to the residences of staff working in the field. These are to be brought back to the operations unit and incarnated or buried deep in the ground. Co-attractants (synthetic kairomones) such as ethyl acetate/ethyl alcohol are known to enhance captures in RPW pheromone traps, but could also significantly increase the cost of an area-wide control programme [54–57].

As regards trap density, in surveillance programmes, set traps along the motorable roads @1 trap for every km. Depending on the pest intensity in mass trapping programmes, 1–4 traps/ha can be adopted [58]. Usually 1 trap/ha is deployed and if more traps are to be set in the field, use service-less trapping options (Attract and Kill; Dry trap-Electrap™) when trap density has to be enhanced beyond 1 trap/ha. Set traps preferably on the ground under the shade with around half of the bucket trap inserted into the soil. Do not place traps directly on young palms. Numbering/geo-referencing of every trap in the field is indispensable for periodic review of the situation in the field and mobilizing resources around traps with high weevil captures. It is important to emphasize that poor bait lure synergy due to sub-standard trapping protocols would end up in the palm smelling better than the trap and attracted weevils getting oriented to the palm instead of entering the trap. This is a very dangerous situation where a poorly maintained trap acts as a catalyst in creating new infestations.

Although the food-baited RPW pheromone trap is most popular, the periodic replacement of the food bait is cumbersome and not sustainable in the long run, especially in area-wide control programmes. In this context, bait and trap free technique of attract and kill has been tested and used to curtail the emerging adult RPW population [59, 60]. Another service-less RPW trapping option that works without the food bait/water is the dry Electrap™ [48]. The cost of incorporating these techniques in an area-wide control programme could be a factor to be considered and needs to be compared with the traditional food-baited pheromone trap before adoption. Large-scale control programmes would stand to benefit if smart traps capable of recording and transmitting weevil capture data on a 24×7 basis are developed. In this context, Potamitis et al. [61] and Aldhryhim and Al-Ayedh [62] have developed and tested smart traps for RPW, but these need advancement for large-scale deployment in the field. An ideal RPW pheromone trap would be the one that does not need servicing and automatically transmits weevil capture data on a 24×7 basis to the operations control unit.

5.3 Chemical treatments

Preventive and curative chemical treatments are essential for the efficient management of RPW [27, 63]. Fajardo [64] and Aldawood [65] have described protocols for these treatments in detail.

5.3.1 Preventive treatments

Preventive chemical treatments are often abused and deployed unnecessarily on a calendar basis resulting in drawbacks associated with these treatments, such as pest developing resistance to chemical pesticides, residues of the chemicals in the fruit (dates/coconut) and resurgence of secondary pests and contamination of the environment. Recent reports suggest that RPW is developing resistance to the several insecticides [66, 67]. Overdependence on pesticide application in date plantations has resulted in the residue level of certain insecticides and acaricides being higher than the maximum residue levels permitted in dates [68]. Preventive treatments should only be carried out in farms with high weevil activity as gauged from high infestation and the removal of infested palms/high trap captures/high seasonal activity in the Middle East during March–May and September–October. It is imperative to treat all fresh wounds on palm immediately after the frond and offshoot removal [26, 69–71]. These operations of cleaning the palm could be preferably adopted during the peak winter months when temperatures are low and not conducive for egg hatch and larval development. The commonly used insecticides for preventive treatments are imidacloprid, thiamethoxam, avermectin, abamectin, chlorpyrifos and phosmet. It should be borne in mind that the preventive insecticide treatments are often unnecessary and excessive, which would have negative impact on the environment as a whole.

5.3.2 Curative treatments

The curative insecticidal treatment of RPW-infested palms in the early stage of attack is an integral part of the control strategy. Such palms recover with insecticide treatment [63, 72, 73]. Ferry [74] provided a detailed protocol of mechanically sanitizing palms in the early stage of attack, while Aldawood [65] presented a comprehensive protocol to inject palms with insecticide in the early stage of attack. Although pressure injectors are used to deliver the insecticide solution to the infested site inside the palm, this technique could damage the palm tissue if the pressure exceeds 2 bar. The diffusion method (gallon method) by cleaning the palm around the infested site on the palm, drilling 4–6 slanting holes 20 cm deep at an angle and pouring insecticide solution into each of the holes is simple, cost-effective and safe. Treat the palm again after 15 days. Once the palm recovers and if the infestation site is close to the ground, cover the treated site with soil to facilitate rooting. For ornamental palms, including ornamental date palms, a new injection technique based on the microinfusion of avermectin insecticide allows to protect the palms at a very low cost over a period of 1 year [75]. Commonly used insecticides for curative treatments are the following: imidacloprid, thiamethoxam, avermectin and abamectin. In organic date plantations, the proven plant origin insecticides would have to be used to treat RPW-infested palms. Several pressure injectors are available in the market which should be used with extreme caution (not >2 bar pressure), to avoid rupture of palm tissue that can lead to death of the treated palm. Ferry and Gomez [63] recommend that only a limited number of stem injections may be carried out in ornamental palms while prohibiting stem injection on a preventive basis in palms grown as food crops.

5.4 Removal of severely infested palms

Even in well-managed RPW-IPM programmes, a certain percentage of the infestations detected is in the advanced stage of attack, where such palms exhibit large tissue damage often harbouring adult weevils with overlapping generations of the pest and have to be removed (eradicated). Such palms disperse adult weevils in the field that result in new infestations which reverse the achievements made in controlling the pest. It is of utmost importance to detect infestations before adult weevils emerge and disperse. Al-Shawaf et al. [76] validated an area-wide RPW IPM programme in the Al-Ahsa oasis of Saudi Arabia using palm removal (eradication) data as a criterion to judge the success of the field operations. Palm eradication levels above 20% of the infested palms are not desirable, which reveals that the pest is proliferating and control tactics need adjustments.

Vidyasagar [77] outlined a detailed protocol for the safe removal and disposal of severely infested palms. The protocol broadly recommends to the following:

1. Identify the badly infested or damaged palms and mark all such palms with a distinct colour tape or spray paint, or a specific number of straps. Infestation due to RPW could be considered severe in date palms if more than 30% of the trunk tissue is damaged at the infestation site.

2. Initiate the removal process as soon as possible. Otherwise, the adults from these infested palms will make their way to healthy palms in the vicinity, making the task much more difficult.

3. As a prophylactic measure, soak, drench or shower the palm crown and also the trunk and bole regions with a recommended pesticide.

In many countries, shredding machines are used to destroy the severely infested palms at another site, where the severely infested palms are cut into logs and the palm pieces (fronds and trunk) are transported to the shredding site.

Utmost care needs to be taken that there are no escapes of the adult weevils during transportation, for which the palm pieces are to be wrapped in plastic wrapping that is sealed with a tape or transported using enclosed trucks. This is a very expensive process and therefore in-situ (on farm site) removal and disposal of severely infested palm tissue by cutting into small pieces (20 × 10 cm) and soaking with insecticide is recommended [75]. In some countries, the removal of severely infested palms is outsourced to private agencies. Here, bureaucratic procedures in issuing work orders to contractors often result in delay which in turn leads to the spread of the weevil.

Abandoned and neglected plantations also harbour the pest and have to be closely monitored for incidence of RPW by intensive inspection campaigns and installing monitor traps. Farmer cooperation to assist in tackling the pest in neglected gardens should also be sought through persistent awareness programmes. The technique of attract and kill is suited for such plantations. If the plantation is dry with no irrigation and the palm tissue is hard, in all probability, RPW will not prefer such a garden. If there is no pest in an abandoned plantation, these palms should not be removed as palm volatiles emitted during the removal process may attract the pest.

5.5 Validating the control programme

In an area-wide RPW IPM programme, the judicious use of resources (men and material) is vital. Often scarce labour and resources have to be used where most required and the control strategy has to be adjusted by providing resources where most needed. In this context, data on weevil captures in traps, infestation reports and removal of severely infested palms could be used to gauge the situation. Faleiro [26] proposed an assumed action threshold of 1% infestation in large plantations. He developed sequential sampling plans to accurately assess the pest status in coconut and date palm [78, 79] based on infestation reports where in the action threshold, the aggregation index of RPW and the risk of making the wrong decision are built into the plan. The sampling plans are efficient tools in decision-making, particularly at very low and high levels of infestation and can be used to assess the performance of RPW- IPM programmes that are in progress. Al-Shawaf et al. [76] analysed the monthly trap capture data, the infestation reports on the removal of severely infested palms and the above sampling plan [79] to categorize infestation in 15 operational areas (4000 ha) in the Al-Ahsa date palm oasis of Saudi Arabia for a period of 6 months between March and September 2011. They found that the IPM strategy adopted had the desired impact in the East of the oasis, but needed minor adjustments in the centre and called for major reinforcement in the North of the Al-Ahsa oasis. Similarly, Hoddle et al. [80] assessed the impact of pheromone trapping, pesticide applications and eradication of the infested date palms for a period of 5 years between 2007 and 2012, for RPW management in 1104 ha in Al Ghowaybah, of the Al-Ahsa oasis in Saudi Arabia. They concluded that the enhanced RPW management programme that commenced in 2009 had a significant impact against the pest.

GIS-based models can also be developed to validate the strategy at periodic intervals based on trap captures and infestation reports [32, 33]. This helps to judiciously use the resources where most required. FAO has proposed a real-time database and a web portal for the management of RPW at the local, national and NENA region. Furthermore, a mobile app for android and iOS smart phones to record geo-referenced data at the field location on a standard form needs to be developed. FAO has initiated the process to validate the *SusaHamra* app to assist farmers in better monitoring and managing the RPW. A global platform is being established for mapping field data and analytics for better decision-making. Furthermore, remote sensing is being combined with artificial intelligence to map palm trees for the improved monitoring of RPW spread at both the regional and global levels ([34, 35]; http://www.fao.org/news/story/en/item/1184673/icode/).

At *level-2*, the RPW-IPM strategy is to be supported by a robust plant quarantine/phytosanitation regime.

5.6 Quarantine/phytosanitation

Over two decades ago, Abraham et al., [27] first highlighted the importance of quarantine treatments to check the spread of RPW through the infested date palm offshoots and recommended to dip the bole of offshoots before transportation in 0.1% chlorpyrophos for 10 min. In Egypt, quarantine and certification programmes recommend zero tolerance for RPW to block the spread of the pest to secondary foci within planted acreage [81]. Later Faleiro [27] proposed to stop movement of planting material from infested plantations within the country and from one country to another. Wherever this was not possible, it was proposed to implement strict pre- and post-entry quarantine regimes, wherein only pest-free and certified planting material can be transported. Besides date palm offshoots transported for farming, the movement of large palms intended for ornamental gardening contributes largely to the spread of this deadly pest [28]. Hence, it is of utmost importance to keep a strict watch on the movement of planting material (offshoots/palms) for both farming and landscape gardening so that only a treated and pest-free material is allowed to be transported within national boundaries [26, 37]. Al-Shawaf et al. [82] recommended to dip date palm offshoots in 0.004% Fipronil for 30 min before transporting to ensure the complete mortality of the hidden larval stages, if any, and complete certification and transport of the treated offshoots to the new planting site within 72 h of treatment. Studies carried out in Spain in *P. canariensis* suggest that a dose of 1.14 g aluminum phosphide/m^3 for 3 days is enough to kill all the stages of RPW in an infested palm tree, and is recommended as a quarantine protocol provided the dose is not phytotoxic to the treated palms [83]. A sound protocol for the treatment of large palms for ornamental gardening needs to be developed.

Although regulations/decrees to regulate the movement of palms for planting exist in several countries, implementing the decrees in letter and spirit is often lacking [37, 84]. In this context, some European Union (EU) guidelines that could be useful are the following: delimitation of survey and demarcated areas, three monthly official inspections of palm nurseries, annual crop declaration, application of phytosanitary treatments, registration of planting material movement and use of plant passport to monitor the trade of palms. Developing certified palm propagation programmes (certified seed) through tissue culture would go a long way ensuring the propagation of a pest-free material. Recently, Chouibani [11] in the FAO guidelines on RPW management outlined detailed phytosanitary protocols for the movement of palms in context of stopping the spread of RPW within national and international borders and proposed to identify the infested zone where the presence of RPW is confirmed and also a buffer zone extending at least 10 km beyond the boundary of the infested zone. A strict vigil is to be maintained on the movement of palms and plant nurseries within the demarcated area. The demarcated area will be declared free from RPW if, during the three consecutive years, RPW has not been detected. Chouibani [11] further recommends that ornamental palm trees originating from recognized nurseries should be imported. Nurseries should be authorized, certified, mapped and regularly inspected by the NPPO of the exporting country.

As regards date palm, only those propagated in vitro in test tubes by laboratories that are officially certified to propagate such materials should be imported.

The RPW-IPM strategy at *level-3* outlines the prospect of palm tolerance to RPW and encourages the use of biological control to combat the pest, besides emphasizing the importance of adopting good agronomic practices.

5.7 Palm resistance to RPW

In perineal crops, such as palms, farmers prefer to cultivate commercial cultivars that are well established and popular in the area/region. These cultivars/varieties are often the most susceptible to RPW. Host plant resistance has not been fully studied and exploited with regard to RPW in spite of some preliminary research that has characterized palm cultivars in terms of tolerance/susceptibility to RPW [20, 85, 86]. The screening techniques to identify resistant RPW cultivars and parental material for use in breeding programmes need to be developed. The molecular markers-assisted breeding programme for the development of RPW resistant cultivars is another avenue that could be pursued. Advanced molecular techniques such as RNAi could hasten the utilization of host plant resistance against RPW [70].

5.8 Agro-techniques and RPW management

Palm density, irrigation methods and protection of tissue immediately after the frond and offshoot removal are probably the most important agro-techniques related to RPW management [87–89].

5.8.1 Palm density

Traditional date palm plantations have palms planted at a close spacing, often restricting sunlight penetration resulting in the build up of in-groove humidity conducive for the development of RPW [90]. Dense planting coupled with flood irrigation and inadequate drainage accelerates the build up of in-groove humidity. In this context, it is recommended to plant new plantations at the recommended spacing to allow sunlight penetration.

5.8.2 Impact of irrigation

In flood-irrigated palms, often, the water touches the trunk at the ground, which encourages adult weevils to oviposit in the collar region of such palms, resulting in new infestations. It is therefore recommended that drip irrigation systems be installed as a precaution for RPW control [87]. Care should be taken to see that palms receive adequate irrigation water as the scarcity of water may result in the incidence of the date palm stem borer *Jebusaea hammerschmidti* [91].

5.8.3 Protecting fresh wounds

Adult weevils are attracted to palm volatiles emitted from the fresh wounds of palm tissue after the frond and offshoot removal. This calls for the immediate treatment of wounds on the palm to mask the emitting palm volatiles and avoid the gravid female weevil getting attracted to these sites for oviposition. As a precaution, these pruning activities could be carried out during winter when temperatures are low which adversely impacts the egg hatch and larval development [22]. No oviposition was observed for females kept below 15°C [92].

5.9 Biological control

Biological control is an integral part of several sustainable IPM programmes the world over. Biocontrol could play a significant role in augmenting the existing pheromone-based RPW-IPM strategy. Conventional control measures against RPW such as mass pheromone trapping and chemical treatments for preventive and curative purposes have not given the desired level of control. A wide range of RPW natural enemies, viz. insects, bacteria, fungi, viruses, yeasts, entomopathogenic nematodes (EPN) and birds have been reported from several countries [26, 93]. Mazza et al. [94] enlisted more than 50 biological control agents against the *Rhynchophorus* group of weevils. Al-Deeb et al. [95] reported the presence of phoretic mites of the genera *Uropoda*, *Uroobovella* and *Curculanoetus* on RPW in UAE. Whether these mites cause a pathological effect on the weevil need to be investigated. In this respect, it has been found that an unidentified species of phoretic mite could cause high mortality among RPW pupae, as well as attenuation of adult weevils (El-Shafie, unpublished data) (**Figure 7**). Yasin et al. [96] reviewed the potential role of potential microbial agents in the control of RPW and concluded that strains of the fungi *Beauveria bassiana* and *Metarhizium anisoplia*e, isolated from naturally infected RPW, could contribute to biological control of this pest. There are several publications to show the efficacy of entomopathogenic nematodes (EPN) and the entomopathogenic fungi (EPF) in the laboratory and semi-field assays [96–101].

Laboratory and semi-field cage studies showed the possibility of infecting RPW adults with *B. bassiana* using pheromone traps [100]. Reports from Spain suggest that the EPN, *Steinernema* sp. [99], and the EPF, *B. bassiana* [102], are promising in the field. The role of parasitic tachinid flies from South America against the *Rhynchophorus* group of weevils could also be another potential avenue to explore [103, 104]. The current RPW-IPM programme could be significantly strengthened if the known biological control agents could be delivered to the target site and also sustained in the field.

At *level-4*, the strategy mainly visualizes capacity building, robust extension programmes to sensitize all stakeholders and farmer participation in the control programme.

Figure 7.
Unidentified phoretic mite on RPW adult (left) and pupa (right) (Photo: Hamadttu A. F. El-Shafie).

5.10 Capacity building and extension programmes

Building the capacity of farmers, home owners and officials is an essential component of the RPW control programme [37]. Dissemination of the latest information on RPW-IPM among all the stakeholders through the electronic and print media can go a long way in strengthening the area-wide RPW-IPM strategy.

5.11 Farmer participation in RPW control

In several countries, an extensive state support is provided while implementing the RPW-IPM programme, keeping the farmers' involvement either bare minimum or almost nil [37]. This makes it very difficult for the authorities to implement the programme. Closed farms, for example, often develop as breeding sites for RPW, diluting the success achieved in controlling the pest in the vicinity of such plantations/home grooves [27]. Farmers' participation and cooperation are vital for any IPM programmes to succeed [105]. Abdedaiem et al. [106] emphasized the need to take up socio-economic studies for better understanding and improving the farmers' involvement in the RPW control programme.

The spread of the pest in vast stretches of date plantations has resulted in outsourcing area-wide RPW-IPM programmes to private companies in some countries. This needs careful planning, implementation and supervision. Here, the lack of experienced staff to oversee operations in the field is a major constraint. Government authorities need to efficiently supervise, monitor and evaluate the control programme implemented by the private company, on a regular basis. Delay in providing necessary inputs (pheromones, insecticides, etc.) where the RPW control is entrusted to private agencies is another major concern. Often there is a delay in finalizing the tender/quotation for the subsequent period before the expiry of the on-going tender resulting in the stoppage of field operations. Any break in the control operations will result in the proliferation and spread of the pest.

6. Challenges of RPW management and future prospects

As a prelude to the scientific consultation and high level meeting on RPW management held in Rome during March 2017, FAO prepared a base document ([1]; http://www.fao.org/3/a-ms664e.pdf) on the current situation of RPW in the NENA region where in the present management practices, challenges/weaknesses and available research and technologies for its improvement are outlined. There are several gaps and challenges in the core components of the current RPW-IPM strategy. The current RPW-IPM programmes based on pheromone trapping and other techniques have been implemented with limited success. Gaps and challenges in almost all the components of the strategy, particularly with regard to early detection of the pest, developing and implementing phytosanitary measures, limited efficacy of biological control agents in the field, lack of farmer participation in the programmes and scarcity of data on socio-economic issues among several other factors have made RPW control and eradication extremely difficult. On the positive side, the pest has been eradicated in the Canary Islands and is approaching eradication in Mauritania. Eradication has also been obtained in various oasis, but new introductions of infested palms have reversed the success achieved [2].

At level-1 of the strategy, the main challenge in early detection is to provide the farmer with a low cost and efficient detection device. There are several laboratories all over the world that have worked on advanced techniques such as detecting chemical signatures, acoustic detection, use of infrared cameras, thermal imaging, satellite imaging/IoT, etc. [39–43]. Meanwhile, the visual inspection of palms to detect infested palms will continue to stay to detect RPW infested palms.

With regard to pheromone trapping, the need to regularly service the food-baited traps to change the food bait, water and record weevil captures is cumbersome and is the main constraint. Trap and bait-free trapping (attract and kill) and dry trapping (Electrap™) have addressed the challenge to some extent although data collection on weevil captures continues to be a gap that needs to be addressed. Ideally a dry trap that automatically records and transmits weevil capture data is a prospect for the future. Besides attract and kill, other semiochemical mediated control methods against RPW such as 'push-pull', involving the use of repellents and attractants [107], and 'attract and infect', involving the spread of biological control agents (EPFs) using pheromone traps [100], needs to be refined and worked upon.

As regards chemical treatments, there is an overdependence on the use of chemical insecticides for both preventive and curative treatments. Research on the efficacy of natural insecticides against RPW needs to be enhanced so that these can be encouraged and incorporated in the control strategy. It is observed that in well-managed plantations, the preventive insecticidal treatments on a regular calendar basis are not required. Furthermore, for curative treatments are pressure injectors really required? These are expensive and need to be operated under the supervision of trained personnel due to the possibility of rupturing the palm tissue leading to death of the palm if the pressure exceeds 2 bar. Instead the methodology for mechanical sanitization coupled with 'drill and inject' using the simple diffusion technique has to be standardized.

The removal and safe disposal of severely infested palms is also another huge challenge. In several countries, this aspect of the strategy is constrained by the use of costly shredding machines that need trained personnel to operate. Besides, there is a danger of the weevils escaping during transportation of the eradicated palm to the shredder outside the farm to the shredding site. In this context, Ferry [74] recommends the processing/destruction of severely infested palms right at the farm itself. The possibility of using small portable shredders needs to be looked into.

Area-wide RPW-IPM programmes generate a huge amount of data that needs to be collected, processed and analysed. Periodic validation/performance analysis of the control programme is vital to have a grip of the situation and use the resources judiciously and is an important challenge that has to be met. Any meaningful validation of the control programme calls for meticulous record keeping, particularly of the weevil captures in the traps and the number and location of palms infested, palms treated and removed (eradicated). Record keeping is to be facilitated by having field maps of the operational areas wherein the position of the traps and infested palms can be plotted. Numbering of each trap in the field is vital in addition to geo-referencing all the traps and infestations reported. In the absence of professional GIS specialists in the programme, spatial and temporal spread sheets can be prepared periodically by plotting on the maps (use different colours) for weevil captures in traps and infested palms detected. High weevil captures or removal of too many infested palms indicates that the strategy is not doing well and needs adjustments.

Systematic geographic information system (GIS)-linked data collection is indispensable, e.g., by using a large number of mapped (traps and weevil catches), the temporal as well as geographical changes in pest distribution can be monitored and infested palms detected by various means can be localized [41]. The future prospect for data collection, compiling and analysing in area-wide RPW-IPM programmes would be to develop and validate mobile apps for smartphones to record geo-referenced data at the field location on a standard form. Recently, FAO has initiated the process to validate the SusaHamra app to assist farmers in better monitoring and managing the RPW. A global platform is being established for mapping field data and analytics for better decision-making.

At level-2 of the strategy, phytosanitation/quarantine is very important. The main gaps and challenges here are: (i) national/regional phytosanitary/quarantine regulations against RPW are not adequately implemented; (ii) treatment protocols to treat palms prior to transportation and also after arrival at destination are not consistent; (iii) implementation of the regulations is weak due to insufficient staff that is often not trained; and (iv) certified planting material is difficult to get [2, 37]. Recently, Chouibani [11] has addressed these concerns in the FAO guidelines on RPW management. Although chemical protocols available for quarantine purposes are available for date palm offshoots, the need to develop an effective treatment protocol to treat and sanitize large palms for ornamental gardening before transportation has to be addressed on priority. It is the responsibility of respective NPPOs to ensure implementation of the phytosanitary measures against RPW.

The lack of biological control agents against RPW that are field worthy is another major concern. Although there are known biological control agents for RPW, delivery to the target site inside the palm and sustainability of these agents needs to be addressed. Furthermore, the importance of adopting the best agro-techniques related to palm density, frond and offshoot removal and irrigation practices is underestimated. There is scope to conduct research on these aspects to quantify the relationship between these factors and the incidence and severity of RPW. Host plant resistance against RPW is also not very well understood and offers a whole new area of exploration where the traditional plant breeding techniques coupled with advanced molecular based breeding techniques could be used to induce resistance against RPW in the popular palm cultivars. The entire genome of the date palm cultivar 'Khalas' has been sequenced [108, 109]. This could facilitate the integration of genetic engineering techniques into date palm breeding programmes that provide mechanisms to overcome the current constraints to conventional breeding in date palm and help incorporate desirable traits of yield, quality and resistance to abiotic and biotic stresses in date palm [110].

In several countries, the RPW-IPM programme is implemented largely by the state where farmers' participation is minimum or none. This impedes the performance and success of the strategy. Mechanisms need to be devised to increase farmers' participation in RPW-IPM programmes. In this context, Abdedaiem et al., [106] emphasized the need to take up socio-economic studies for enhancing farmers' involvement in the RPW control programme.

During the last few years, there are several publications that have generated data on the cutting-edge molecular aspects of RPW involving RNAi, gene expression, etc. [111–114]. Results of these studies need to be exploited in a way that the control strategy stands to benefit from such basic research.

Faleiro et al. [2] report that in recent years, a large number of new RPW-IPM tools (detectors, surveillance drones, pesticides, palm injectors, semiochemicals, biological control agents, palm shredders, microwave treatment devices, etc.) are available in the market. These IPM tools need proper testing and validation at the national and regional levels so that only field worthy technologies that are not costly and easy to use are made available to the farmers.

7. Conclusions

The red palm weevil remains to be the invasive key pest threatening palms survival around the world. International trading and transportation of infested planting material (palms) for plantations and landscape purposes are the main introduction pathways into new non-invaded areas. Current management strategies against RPW depend on monitoring and mass trapping using pheromones, agronomic and phytosanitary measures and to some extent biological control. Capacity building and quarantine measures are also among the RPW-IPM components. Despite major global efforts to combat the weevil, many gaps and challenges, in management strategies, need to be addressed. Such challenges include early detection of infestation, optimization of pheromone-baited traps, removal of highly infested palms, overdependence on the use of insecticides and participation of farmers in the control efforts. Future prospects of RPW management may include validation of management programmes, testing of high-tech technologies for practical field application and the use of RNAi technology in management programmes. It can be concluded that managing RPW in the field is not an easy task but with adequate resources, appropriate interventions supported by good coordination, planning and financial resources, the pest can be effectively controlled with the current technologies.

Acknowledgements

The authors acknowledge and appreciate the financial support of this work provided by the Date Palm Research Center of Excellence, King Faisal University, through funding the research project number DPRC-9-2018.

Author details

Hamadttu Abdel Farag El-Shafie[1]* and Jose Romeno Faleiro[2]

1 Date Palm Research Center of Excellence, King Faisal University, Al-Ahsa, Kingdom of Saudi Arabia

2 Ex-Indian Council of Agricultural Research, Goa, India

*Address all correspondence to: elshafie62@yahoo.com

References

[1] FAO. Current situation of management practices, challenges/ weaknesses and available research and technologies for its improvement. Current Situation of Red Palm Weevil in the NENA Region. 2017. Available from: http://www.fao.org/3/a-ms664e.pdf [Accessed: 14 May 2020]

[2] Faleiro JR, Ferry M, Yaseen T, Al-Dobai S. Overview of the gaps, challenges and prospects of red palm weevil management. In: Presented at the International Scientific Meeting on 'Innovative and Sustainable Approaches to Control the Red Palm Weevil', CIHEAM Bari; 23-25 October 2018; Organized by FAO and CIHEAM Bari, Italy [Published in Arab Journal of Plant Protection. 2019;**37**(2):170-177]. 2018. DOI: 10.22268/AJPP-037.2.170177

[3] Fiaboe KKM, Peterson AT, Kairo MTK, Roda AL. Predicting the potential worldwide distribution of the red palm weevil *Rhynchophorus ferrugineus* (Olivier) (Coleoptera: Curculionidae) using ecological niche modelling. Florida Entomologist. 2012;**95**:559-673

[4] Roda A, Kairo M, Damian T, Franken F, Heidweiller K, Johanns C, et al. Red palm weevil (*Rhynchophorus ferrugineus*), an invasive pest recently found in the Caribbean that threatens the region. Bulletin OEPP (EPPO Bulletin). 2011;**41**(2):116-151. DOI: 10.1111/j.1365-2338.2011.02446.x

[5] CDFA. Red Palm Weevil, Worst Known Pest of Palm Trees Detected in Laguna Beach. California Department of Food and Agriculture (CDFA); 2010. Press Release # 10-061

[6] Rugman-Jones PF, Hoddle CD, Hoddle MS, Stouthamer R. The lesser of two weevils: Molecular genetics of pest palm weevil populations confirm *Rhynchophorus vulneratus* (panzer 1798) as a valid species distinct from *R. ferrugineus* (Olivier 1790), and reveal the global extent of both. PLoS One. 2013;**8**(10):e78379. DOI: 10.1371/ journal.pone.0078379

[7] EPPO. *Rhynchophorus ferrugineus*. EPPO datasheets on pests recommended for regulation. 2020. Available from: https://gd.eppo.int [Accessed: 15 May 2020]

[8] Nirula KK. Investigation on the pests of coconut palm, part-IV. *Rhynchophorus ferrugineus*. Indian Coconut Journal. 1956;**9**:229-247

[9] Anonymous. Save Algarve palms. 2013. Available from: http://www.savealgarvepalms.com/en/weevil-facts/host-palm-trees. [Accessed: 14 May 2020]

[10] Giblin-Davis RM, Faleiro JR, Jacas JA, Peña JE, Vidyasagar PSPV. Coleoptera: Biology and management of the red palm weevil, *Rhynchophorus ferrugineus*. In: Peña JE, editor. Potential Invasive Pests of Agricultural Crop Species. Wallingford, UK: CABI; 2013. pp. 1-34

[11] Chouibani M. Guidelines on phytosanitary inspections. In: Elkakhy M, Faleiro JR, editors. Red Palm Weevil: Guidelines on Management Practices. Rome: FAO; 2020. DOI: 10.4060/ca7703en

[12] Faleiro JR, Ashok Kumar J, Rangnekar PA. Spatial distribution of red palm weevil *Rhynchophorus ferrugineus* Oliv. (Coleoptera: Curculionidae) in coconut plantations. Crop Protection. 2002;**21**:171-176

[13] Ávalos JA, Martí-Campoy A, Soto A. Study of the flying ability of *Rhynchophorus ferrugineus* (Coleoptera: Dryopthoridae) adults using a computerized flight mill. Bulletin

of Entomological Research. 2014;**104**:462-470

[14] Hoddle MS, Hoddle CD, Faleiro JR, El-Shafie HAF, Jeske DR, Sallam AA. How far can the red palm weevil (Coleoptera: Curculionidae) fly? Computerized flight mill studies with field-captured weevils. Journal of Economic Entomology. 2015;**1-11**(2015). DOI: 10.1093/jee/tov240

[15] Al-Ayedh H. Red palm weevil biology. In: Elkakhy M, Faleiro JR, editors. Red Palm Weevil: Guidelines on Management Practices. Rome: FAO; 2020. DOI: 10.4060/ca7703en

[16] Wattanapongsiri A. A Revision of the Genera *Rhynchophorus* and *Dynamis* (Coleoptera: Cuculionidae). Vol. 1. Bangkok, Thailand: Department of Agriculture Science Bulletin; 1966. pp. 328

[17] Avand Faghih A. The biology of red palm weevil, *Rhynchophorus ferrugineus* Oliv. (Coleoptera: Curculionidae) in Saravan region (Sistan and Balouchistan Province, Iran). Applied Entomology and Phytopathology. 1996;**63**:16-18

[18] Jaya S, Suresh T, Sobhitha Rani RS, Sreekumar S. Evidence of seven larval instars in *Rhynchophorus ferrugineus* reared on sugarcane. Journal of Entomological Research. 2000;**24**:27-31

[19] Abraham VA, Faleiro JR, Nair CPR, Nair SS. Present management technologies for red palm weevil *Rhynchophorus ferrugineus* Olivier (Coleoptera: Curculionidae) in palms and future thrusts. Pest Management in Horticultural Ecosystems. 2002;**8**:69-82

[20] Al-Ayedh H. Evaluation of date palm cultivars for rearing the red date palm weevil, *Rhynchophorus ferrugineus* (Coleoptera: Curculionidae). Florida Entomologist. 2008;**91**:353-358

[21] Yu R, Leung P. Optimal Pest management: A reproductive pollutant perspective. International Journal of Pest Management. 2006;**52**:3155-3166

[22] Dembilio Ó, Jacas JA. Basic bio-ecological parameters of the invasive red palm weevil, *Rhynchophorus ferrugineus* (Coleoptera: Curculionidae), in *Phoenix canariensis* under Mediterranean climate. Bulletin of Entomological Research. 2011;**101**:153-163

[23] El-Shafie HAF, Faleiro JR, Abo-El-Saad MM, Aleid SM. A meridic diet for laboratory rearing of red palm weevil, *Rhynchophorus ferrugineus* (Coleoptera: Curculionidae). Scientific Research and Essays. 2013;**8**(39):1924-1932

[24] Abraham VA, Kurian C. An integrated approach to the control *Rhynchophorus ferrugineus,* the red weevil of coconut palm. In: Proceedings, 4th Session of the FAO Technical Working Party on Coconut Production, Protection and Processing; 14-25 September; Kingston, Jamaica. 1975

[25] Al-Ayedh HY, Al Dhafer HM. Does *Oryctes elegans* (Coleoptera: Scarabaeidae) abundance determine future abundance of *Rhynchophorus ferrugineus* (Coleoptera: Rhynchophoridae) in the date palms of Saudi Arabia? African Entomology. 2015;**23**(1):43-47

[26] Faleiro JR. A review of the issues and management of the red palm weevil *Rhynchophorus ferrugineus* (Coleoptera: Rhynchophoridae) in coconut and date palm during the last one hundred years. International Journal of Tropical Insect Science. 2006;**26**:135-154

[27] Abraham VA, Al-Shuaibi MA, Faleiro JR, Abozuhairah RA, Vidyasagar PSPV. An integrated management approach for red palm weevil, *Rhynchophorus ferrugineus* Oliv., a key pest of date palm in the Middle

East. Sultan Qaboos University Journal for Scientific Research (Agricultural Sciences). 1998;**3**:77-83

[28] Faleiro JR, Ben Abdallah A, El Bellaj M, Al-Ajlan AM, Oihabi A. Threat of red palm weevil, *Rhynchophorus ferrugineus* (Olivier) to date plantations of the Maghreb region in North Africa. Arab Journal of Plant Protection. 2012;**30**:274-280

[29] Jaques JA. Guidelines on visual inspection for early detection of red palm weevil in Canary Island palm (*Phoenix canariensis*). In: Elkakhy M, Faleiro JR, editors. Red Palm Weevil: Guidelines on Management Practices. Rome: FAO; 2020. DOI: 10.4060/ca7703en

[30] Brun L, Kamal W, Soliman A. The date palm in Egypt: A traditional cash crop threatened by the red palm weevil. In: Proceedings of the 1st International Workshop on Red Palm Weevil; 28-29 November 2005; IVIA, Valencia, Spain. 2006. p. 180

[31] Massoud AM, Faleiro JR, El-Saad MA, Sultan E. Geographic information system used for assessing the red palm weevil *Rhynchophorus ferrugineus* (Olivier) in date palm oasis of Al-Hassa, Saudi Arabia. Journal of Plant Protection Research. 2011;**51**(3):234-239

[32] Massoud MA, Sallam AA, Faleiro JR, Al-Abdan S. Geographic information system-based study to ascertain the spatial and temporal spread of red palm weevil *Rhynchophorus ferrugineus* (Coleoptera: Curculionidae) in date plantations. International Journal of Tropical Insect Science. 2012;**32**(2):108-115

[33] Fajardo M, Guerra JA, Barroso L, Morales M, Martín R. Use of GIS (geographical information system) for data analysis in a *Rhynchophorus ferrugineus* eradication program. In: Al-Dobai S, ElKakhy M, Faleiro R, editors. Proceedings of the Scientific Consultation and High-Level Meeting on Red Palm Weevil Management; 29-31 March 2017; Rome, Italy. FAO. 2019. pp. 200 Licence: CC BY-NC-SA 3.0 IGO. ISBN: 978-92-5-130961-2.192p

[34] Yaseen T. RPW trust fund programme for the eradication of RPW. In: Presented at the International Scientific Meeting on 'Innovative and Sustainable Approaches to Control the Red Palm Weevil'; CIHEAM Bari; 23-25 October 2018; Organized by FAO and CIHEAM Bari, Italy. 2018

[35] Cressman K. Red palm weevil monitoring and early warning system. In: Presented at the International Scientific Meeting on 'Innovative and Sustainable Approaches to Control the Red Palm Weevil', CIHEAM Bari; 23-25 October 2018; Organized by FAO and CIHEAM Bari, Italy[Published in Arab Journal of Plant Protection. 2019;**37**(2):203-204]. 2018. DOI: 10.22268/AJPP 037.2.203204

[36] Ferry M, AlDobai S, ElKakhy H. The state of art of the control of the red palm weevil. In: Presented at the Sixth International Date Palm Conference; 19-21 March; Organized by Khalifa International Award for Date Palm and Agricultural Innovation. 2018

[37] FAO. In: Al-Dobai S, ElKahkyy M, Faleiro R, editors. Proceedings of the Scientific Consultation and High-Level Meeting on Red Palm Weevil Management; 29-31 March 2017; Rome, Italy. Rome. 2019. pp. 200. Licence: CC BY-NC-SA 3.0 IGO

[38] Vidyasagar PSPV. Guidelines on visual inspection for early detection of red palm weevil in date palm (*Phoenix dactylifera*). In: Elkakhy M, Faleiro JR, editors. Red Palm Weevil: Guidelines on Management Practices. Rome: FAO; 2020. DOI: 10.4060/ca7703en

[39] El-Faki MS, El-Shafie HAF, Al-Hajhoj MR. Potentials for early detection of red palm weevil (Coleoptera: Curculionidae) infested date palm (Arecaceae) using temperature differentials. Canadian Entomologist. 2015. DOI: 10.4039/tce.2015.51

[40] Mankin RW. Towards user-friendly early detection acoustic devices and automated monitoring for red palm weevil management. In: Al-Dobai S, ElKakhy M, Faleiro R, editors. Proceedings of the Scientific Consultation and High-Level Meeting on Red Palm Weevil Management; 29-31 March 2017; Rome, Italy. FAO. 2017, 2019. pp. 200. Licence: CC BY-NC-SA 3.0 IGO. ISBN: 978-92-5-130961-2.192p

[41] Soroker V, Suma P, La Pergola A, Navarro-Llopis V, Vacas S, Cohen Y, et al. Detection of red palm weevil infestation. In: Al-Dobai S, ElKakhy M, Faleiro R, editors. Proceedings of the Scientific Consultation and High-Level Meeting on Red Palm Weevil Management; 29-31 March 2017; Rome, Italy. FAO. 2017, 2019. pp. 200. Licence: CC BY-NC-SA 3.0 IGO. ISBN: 978-92-5-130961-2.192p

[42] Pugliese M, Rettori AA, Martinis R, Al-Rohily K, Al-Maashi A. Devices to detect red palm weevil infestation on palm species. Precision Agriculture. 2018;**19**(6):1049-1061

[43] Ashry I, Mao Y, Al-Fehaid Y, Al-Shawaf A, Al-Bagshi M, Al-Brahim S, et al. Early detection of red palm weevil using distributed optical sensor. Science Report. 2020;**10**:3155. DOI: 10.1038/s41598-020-60171-7

[44] Hallett RH, Gries G, Gries R, Borden JH, Czyzewska E, Oehlschlager AC, et al. Aggregation pheromones of two Asian palm weevils *Rhynchophorus ferrugineus* and *R. vulneratus*. Naturwissenschaften. 1993;**80**:328-331

[45] Hallett RH, Oehlschlager AC, Borden JH. Pheromone trapping protocols for the Asian palm weevil, *Rhynchophorus ferrugineus* (Coleoptera: Curculionidae). International Journal of Pest Management. 1999;**45**:231-237

[46] Vidyasagar PSPV, Hagi M, Abozuhairah RA, Al-Mohanna OE, Al-Saihati AA. Impact of mass pheromone trapping on red palm weevil adult population and infestation level in date palm gardens of Saudi Arabia. Planter. 2000;**76**(891):347-355

[47] Oehlschlager AC. Palm weevils' pheromones: Discovery and use. Journal of Chemical Ecology. 2016;**42**:617-630

[48] Al-Saroj SE, Al-Abdallah AM, Al-Shawaf AM, Al-Dandan I, Al-Abdullah A, Al-Shagag Y, et al. Efficacy of bait free pheromone trap (Electrap™) for management of red palm weevil, *Rhynchophorus ferrugineus* (Olivier) (Coleoptera: Curculionidae). Pest Management in Horticultural Ecosystems. 2017;**23**(1):55-59

[49] Soroker V, Harari A, Faleiro JR. The role of semiochemicals in date pest management. In: Wakil W, Faleiro JR, Miller T, editors. Sustainable Pest Management. Date Palm: Current Status and Emerging Challenges. Switzerland: Springer International Publishing; 2015. pp. 445. ISBN: 978-3-319-24397-9

[50] Vacas S, Abad-Payá M, Primo J, Navarro-Llopis V. Identification of pheromone synergists for *Rhynchophorus ferrugineus* trapping systems from *Phoenix canariensis* palm volatiles. Journal of Agricultural and Food Chemistry. 2014;**62**:6053-6064

[51] El-Shafie HAF, Faleiro JR. Optimizing components of pheromone-baited trap for the management of red palm weevil, *Rhynchophorus ferrugineus* (Coleoptera: Curculionidae) in date palm agroecosystem. Journal of Plant

Diseases and Protection. 2017. DOI: 10.1007/s41348-017-0097-5

[52] Faleiro JR, Al-Shawaf AM. IPM of red palm weevil. In: El Bouhssini M, Faleiro JR, editors. Date Palm Pests and Diseases: Integrated Management Guide 2018. International Centre for Agricultural Research in the Dry Areas (ICARDA); 2018. pp. 179. ISBN 13: 978-92-9127-505-2

[53] Faleiro JR. Guidelines on RPW pheromone trapping with respect to trapdesign, trap density and servicing. In: Elkakhy M, Faleiro JR, editors. Red Palm Weevil: Guidelines on Management Practices. Rome: FAO; 2020. DOI: 10.4060/ca7703en

[54] Oehlschlager AC. Trapping of date palm weevil. In: Proceedings of an FAO Workshop on Date Palm Weevil (*Rhynchophorus ferrugineus*) and its Control; 15-17 December 1998; Cairo, Egypt. 1998

[55] Al-Shagag A, Al-Abbad AH, Al-Dandan AM, Ben Abdallah A, Faleiro JR. Enhancing trapping efficiency of red palm weevil pheromone traps with ethyl acetate. Indian Journal of Plant Protection. 2008;**36**:310-311

[56] Al-Saoud AH. Effect of ethyl acetate and trap colour on weevil captures in red palm weevil *Rhynchophorus ferrugineus* (Coleoptera: Curculionidae) pheromone traps. International Journal of Tropical Insect Science. 2013;**33**(3):202-206

[57] Vacas S, Primo J, Navarro-Llopis V. Advances in the use of trapping systems for *Rhynchophorus ferrugineus* (Coleoptera: Curculionidae): Traps and attractants. Journal of Economic Entomology. 2013;**106**:1739-1746

[58] Faleiro JR, El-Saad MA, Abdul Hadi AH. Pheromone trap density to mass trap *Rhynchophorus ferrugineus* (Coleoptera: Curculionidae/Rhynchophoridae/Dryophthorid) in date plantations of Saudi Arabia. International Journal of Tropical Insect Science. 2011;**31**(1-2):75-77

[59] El-Shafie HAF, Faleiro JR, Al-Abbad AH, Stoltman L, Mafra-Neto A. Bait-free attract and kill technology (hook™ RPW) to suppress red palm weevil, *Rhynchophorus ferrugineus* (Coleoptera: Curculionidae) in date palm. Florida Entomologist. 2011;**94**(4):774-778

[60] Faleiro JR, Al-Shawaf AM, El-Shafie HAF, Pai Raikar S. Studies on service free semiochemical mediated technologies to control red palm weevil *Rhynchophorus ferrugineus* Olivier based on trials in Saudi Arabia and India. In: Presented at the International Scientific Meeting on 'Innovative and Sustainable Approaches to Control the Red Palm Weevil', CIHEAM Bari; 23-25 October 2018; Organized by FAO and CIHEAM Bari, Italy [Published in Arab Journal of Plant Protection. 2019;**37**(2):136-142]. 2018. DOI: 10.22268/AJPP 037.2.203204

[61] Potamitis I, Ganchev T, Kontodimas D. On automatic bioacoustics detection of pests: The cases of *Rhynchophorus ferrugineus* and *Sitophilus oryzae*. Journal of Economic Entomology. 2009;**102**(4):1681-1690

[62] Aldhryhim YN, Al-Ayedh HY. Diel flight activity patterns of the red palm weevil (Coleoptera: Curculionidae) as monitored by smart traps. Florida Entomologist. 2015;**98**(4):1019-1024

[63] Ferry M, Gomez S. Assessment of risks and potential of injection techniques in integrated programs to eradicate the red palm weevil: Review and new perspectives. Fruits. 2014;**69**:143-157

[64] Fajardo M. Guidelines on preventive pesticide treatments (sprays/showers). In: Elkakhy M, Faleiro JR, editors.

Red Palm Weevil: Guidelines on Management Practices. Rome: FAO; 2020. DOI: 10.4060/ca7703en

[65] Aldawood A. Guidelines on curative pesticide treatments (chemical trunk injection). In: Elkakhy M, Faleiro JR, editors. Red Palm Weevil: Guidelines on Management Practices. Rome: FAO; 2020. DOI: 10.4060/ca7703en

[66] Al-Ayedh H, Hussain A, Rizwan-ul-Haq M, Al-Jabr AM. Status of insecticide resistance in field-collected populations of *Rhynchophorus ferrugineus* (Olivier) (Coleoptera: Curculionidae). International Journal of Agriculture and Biology. 2016;**18**:103-110

[67] Wakil W, Yasin M, Qayyum MA, Ghazanfar MU, Al-Sadi AM, Bedford GO, et al. Resistance to commonly used insecticides and phosphine fumigant in red palm weevil, *Rhynchophorus ferrugineus* (Olivier) in Pakistan. PLoS One. 2018;**13**(7):e0192628. DOI: 10.1371/journal.pone.0192628

[68] El-Saeid MH, Al-Dosari SA. Monitoring of pesticide residues in Riyadh dates by SFE, MSE, SFC, and GC techniques. Arabian Journal of Chemistry. 2010;**3**(3):179-186

[69] Dembilio Ó, Riba JM, Gamón M, Jacas JA. Mobility and efficacy of abamectin and imidacloprid against *Rhynchophorus ferrugineus* in *Phoenix canariensis* by different application methods. Pest Management Science. 2015;**71**(8):1091-1098

[70] Al-Dosary NM, Al-Dobai S, Faleiro JR. Review on the management of red palm weevil *Rhynchophorus ferrugineus* Olivier in date palm *Phoenix dactylifera* L. Emirates Journal of Food and Agriculture. 2016;**28**(1):34-44. DOI: 10.9755/ejfa.2015-10-897

[71] Milosavljević I, El-Shafie HAF, Faleiro JR, Hoddle CD, Lewis M, Hoddle MS. Palmageddon: The wasting of ornamental palms by invasive palm weevils, *Rhynchophorus* spp. Journal of Pest Science. 2018;**92**:143-156

[72] Aldawood AN, Alsagan F, Altuwariqi H, Almuteri A, Rasool K. Red palm weevil chemical treatments on date palms in Saudi Arabia: Results of extensive experimentations. In: Colloque méditerranéen sur les ravageurs des palmiers; 16-18 Janvier 2013; Nice, France. 2013

[73] Gomez S, Ferry M. A simple and low cost injection technique to protect efficiently ornamental Phoenix against the red palm weevil during one year. In: Presented at the International Scientific Meeting on 'Innovative and Sustainable Approaches to Control the Red Palm Weevil', CIHEAM Bari; 23-25 October 2018; Organized by FAO and CIHEAM Bari, Italy. 2018. DOI: 10.22268/AJPP 037.2.124129 [Published in Arab Journal of Plant Protection. 2019;**37**(2):124-129]

[74] Ferry. Guidelines on mechanical sanitization of infested palms and removal of severely infested palms. In: Elkakhy M, Faleiro JR, editors. Red Palm Weevil: Guidelines on Management Practices. Rome: FAO; 2020. DOI: 10.4060/ca7703en

[75] Ferry M. Recent advances in insecticide treatments against the red palm weevil. In: Al-Dobai S, ElKakhy M, Faleiro R, editors. 2019: Proceedings of the Scientific Consultation and High-Level Meeting on Red Palm Weevil Management; 29-31 March 2017; Rome, Italy. FAO. 2017. pp. 200. Licence: CC BY-NC-SA 3.0 IGO. ISBN: 978-92-5-130961-2.192p

[76] Al-Shawaf AM, Al-Abdan S, Al-Abbad AH, Ben Abdallah A, Faleiro JR. Validating area-wide management of *Rhynchophorus ferrugineus* (Coleoptera: Curculionidae) in date plantation of Al-Hassa.

Indian Journal of Plant Protection. 2012;**40**(4):255-259

[77] Vidyasagar PSPV. Guidelines on removal and safe disposal of highly infested and damaged palms. In: Elkakhy M, Faleiro JR, editors. Red Palm Weevil: Guidelines on Management Practices. Rome: FAO; 2020. DOI: 10.4060/ca7703en

[78] Faleiro JR, Ashok Kumar J. A rapid decision sampling plan for implementing area-wide management of red palm weevil, *Rhynchophorus ferrugineus*, in coconut plantations of India. Journal of Insect Science. 2008;**8**:15. Available online :insectscience.org/8.15. 9 pp

[79] Faleiro JR, Ben Abdallah A, Ashok Kumar J, Shagagh A, Al-Abdan S. Sequential sampling plan for area-wide management of red palm weevil, *Rhynchophorus ferrugineus* (Olivier) in date plantations of Saudi Arabia. International Journal of Tropical Insect Science. 2010;**30**(3):145-153

[80] Hoddle MS, Al-Abbad AH, El-Shafie HAF, Faleiro JR, Sallam AA, Hoddle CD. Assessing the impact of pheromone trapping, pesticide applications, and eradication of infested date palms for *Rhynchophorus ferrugineus* (Coleoptera: Curculionidae) management in Al Ghowaybah, Saudi Arabia. Crop Protection. 2013;**53**:152-160

[81] Salama HS, Abd-Elgawad MM. Quarantine problems: An analytical approach with special reference to palm weevils and phytonematodes. Archives of Phytopathology and Plant Protection. 2003;**36**:41-46

[82] Al-Shawaf AM, Al-Shagag A, Al-Bagshi M, Al-Saroj S, Al-Bather S, Al-Dandan AM, et al. A quarantine protocol against red palm weevil *Rhynchophorus ferrugineus* (Olivier) (Coleptera: Curculiondae) in date palm. Journal of Plant Protection Research. 2013;**53**(4):409-415

[83] Llácer E, Jacas JA. Efficacy of phosphine as a fumigant against *Rhynchophorus ferrugineus* (Coleoptera: Curculionidae) in palms. Spanish Journal of Agricultural Research. 2010;**8**(3):775-779

[84] Balijepalli SB, Faleiro JR. Is policy paralysis on quarantine issues in the near east and North Africa region leading to the build-up and spread of red palm weevil? In: Presented at the International Scientific Meeting on 'Innovative and Sustainable Approaches to Control the Red Palm Weevil', CIHEAM Bari; 23-25 October 2018; Organized by FAO and CIHEAM Bari, Italy. 2018. DOI: 10.22268/AJPP-037.2.83-88 [Published in Arab Journal of Plant Protection. 2019;**37**(2):83-88]

[85] Dembilio Ó, Jacas JA, Llácer E. Are the palms *Washingtonia filifera* and *Chamaerops humilis* suitable hosts for the red palm weevil, *Rhynchophorus ferrugineus* (Coleoptera: Curculionidae). Journal of Applied Entomology. 2009;**133**:565-567

[86] Faleiro JR, El-Shafie HAF, Ajlan AM, Sallam AA. Screening date palm cultivars for resistance to red palm weevil *Rhynchophorus ferrugineus* (Coleptera: Curculiondae). Florida Entomologist. 2014;**97**(4):1529-1536

[87] Sallam AA, El-Shafie HAF, Al-Abdan S. Influence of farming practices on infestation by red palm weevil *Rhynchophorus ferrugineus* (Olivier) in date palm: A case study. International Research Journal of Agricultural Science and Soil Science. 2012;**2**:370-376

[88] Ben Salah M. Importance of field operations for reducing red palm weevil (RPW) infestation on date palm. In: Presented at the International Scientific Meeting on 'Innovative and Sustainable

Approaches to Control the Red Palm Weevil', CIHEAM Bari; 23-25 October 2018; Organized by FAO and CIHEAM Bari, Italy. 2018. DOI: 10.22268/AJPP-037.2.159-162 [Published in Arab Journal of Plant Protection. 2019;**37**(2):159-162]

[89] Faleiro JR, Aldawood AS. Guidelines on good agronomic practices (including palm density in the field, irrigation, and crop and field sanitation). In: Elkakhy M, Faleiro JR, editors. Red Palm Weevil: Guidelines on Management Practices. Rome: FAO; 2020. DOI: 10.4060/ca7703en

[90] Aldryhim Y, Khalil A. Effect of humidity and soil type on survival and behaviour of red palm weevil *Rhynchophorus ferrugineus* (Oliv.) adults. Sultan Qaboos University Journal for Scientific Research (Agricultural and Marine Sciences). 2003;**8**:87-90

[91] El-Shafie HAF. Biology, ecology and management of the long horn date palm stem borer *Jebusaea hammerschmidti* (Coleoptera: Cerambycidae). Outlooks on Pest Management. 2015:20-23. DOI: 10.1564/v26_feb_06

[92] Dembilio Ó, Tapia GV, Téllez MM, Jacas JA. Lower temperature thresholds for oviposition and egg hatching of the red palm weevil, *Rhynchophorus ferrugineus* (Coleoptera: Curculionidae), in a Mediterranean climate. Bulletin of Entomological Research. 2012;**102**:97-102

[93] Murphy ST, Briscoe BR. The red palm weevil as an alien invasive: Biology and the prospects for biological control as a component of IPM. Biocontrol News and Information. 1999;**20**(1):35-45

[94] Mazza G, Francardi V, Simoni S, Benvenuti C, Cervo R, Faleiro JR, et al. An overview on the natural enemies of *Rhynchophorus* palm weevils, with focus on *R. ferrugineus*. Biological Control. 2014;**77**:83-92

[95] Al-Deeb MA, Bin Muzaffar S, Abuagla AM, Sharif EM. Distribution and abundance of phoretic mites (Astigmata, Mesostigmata) on *Rhynchophorus ferrugineus* (Coleoptera: Curculionidae). Florida Entomologist. 2011;**94**(4):748-755

[96] Yasin M, Wakil W, Ghazanfar MU, Qayyum MA, Tahir M, Bedford GO. Virulence of entomopathogenic fungi *Beauveria bassiana* and *Metarhizium anisopliae* against red palm weevil, *Rhynchophorus ferrugineus* (Olivier). Entomological Research. 2019;**49**(1):3-12

[97] Hanounik SB, Saleh MME, Abuzuairah RA, Alheji M, Aldhahir H, Aljarash Z. Efficacy of entomopathogenic nematodes with antidesiccants in controlling the red palm weevil *Rhynchophorus ferrugineus* on date palm trees. International Journal of Nematology. 2000;**10**:131-134

[98] Ghazavi M, Avand-Faghih A. Isolation of two entomopathogenic fungi on red palm weevil *Rhynchophorus ferrugineus* (Oliv.) (Coleoptera: Curculionidae) in Iran. Applied Entomology and Phytopathology. 2002;**69**:44-45

[99] Dembilio Ó, Quesada-Moraga E, Santiago-Alvarez C, Jacas JA. Biocontrol potential of an indigenous strain of the entomopathogenic fungus *Beauveria bassiana* (Ascomycota; Hypocreales) against the red palm weevil, *Rhynchophorus ferrugineus* (Coleoptera: Curculionidae). Journal of Invertebrate Pathology. 2010;**104**:214-221

[100] Hajjar MJ, Ajlan AM, Al-Ahmad MH. New approach of *Beauveria bassiana* to control the red palm weevil (Coleoptera: Curculionidae) by trapping technique. Journal of Economic Entomology. 2015;**1-8**(2015). DOI: 10.1093/jee/tou055

[101] Al-Keridis LA, Gaber NM, Aldawood AS. Pathogenicity of Saudi Arabian fungal isolates against egg and larval stages of *Rhynchophorus ferrugineus* under laboratory conditions. International Journal of Tropical Insect Science. 2020. DOI: 10.1007/s42690-020-00141-8

[102] Güerri-Agulló B, López-Follana R, Asensio L, Barranco P, Lopez-Llorca LV. Use of a solid formulation of *Beauveria bassiana* for biocontrol of the red palm weevil (*Rhynchophorus ferrugineus*) (Coleoptera: Dryophthoridae) under field conditions in SE Spain. Florida Entomologist. 2011;**94**(4):737-747

[103] Moura JIL, Mariau D, Delabie JHC. Efficacy of *Paratheresia menezesi* Townsend (Diptera: Tachinidae) for natural biological control of *Rhynchophorus palmarum* (L.) (Coleoptera: Curculionidae). Oléagineux. 1993;**48**(5):219-222

[104] Moura JIL, Toma R, Sgrillo RB, Delabie JHC. Eficiência do parasitismo natural por *Billaea rhynchophorae* (Blanchard) (Diptera: Tachinidae) Para o controle de *Rhynchophorus palmarum* (L.) (Coleoptera: Curculionidae). Neotropical Entomol. ogy. 2006;**35**:273-274

[105] Ju RT, Wang F, Wan FH, Li B. Effect of host plants on development and reproduction of *Rhynchophorus ferrugineus* (Olivier) (Coleoptera: Curculionidae). Journal of Pest Science. 2010;**84**:33-39

[106] Abdedaiem S, Nasr N, Ferry M. Socio-economic studies and approaches for farmer involvement in the RPW control program. In: Al-Dobai S, Elkakhy M, Faleiro R, editors. Proceedings of the Scientific Consultation and High-Level Meeting on Red Palm Weevil Management; 29-31 March 2017; Rome, Italy. Rome: FAO. 2017, 2019. pp. 200. Licence: CC BY-NC-SA 3.0 IGO

[107] Guarino S, Peri E, Bue PL, Germanà MP, Colazza S, Anshelevich L, et al. Assessment of synthetic chemicals for disruption of *Rhynchophorus ferrugineus* response to attractant-baited traps in an urban environment. Phytoparasitica. 2013;**41**:79-88

[108] Al-Dous EK, Binu G, Al-Mahmoud ME, Al-Jaber MY, Wang H, Salameh YM, et al. De novo genome sequencing and comparative genomics of date palm (*Phoenix dactylifera*). Nature Biotechnology. 2011;**29**(6):521-527

[109] Al-Mssallem IS, Hu S, Zhang X, Lin Q, Liu W, Tan J, et al. Genome sequence of the date palm *Phoenix dactylifera* L. Nature Communications. 2013;**4**:2274. DOI: 10.1038/ncomms3274. Available from: www.nature.com/naturecommunications. 1-9 pp

[110] El-Hadrami A, Al-Khayri JM. Socioeconomic and traditional importance of date palm. Emirates Journal of Food and Agriculture. 2012;**24**(5):371-385

[111] Soffan A, Antony B, Abdelazim M, Shukla P, Witjaksono W, Aldosari SA, et al. Silencing the olfactory co-receptor RferOrco reduces the response to pheromones in the red palm weevil, *Rhynchophorus ferrugineus*. PLoS One. 2016;**11**(9):e0162203. DOI: 10.1371/journal.pone.0162203

[112] Antony B, Soffan A, Jakše J, Abdelazim MM, Aldosari SA, Aldawood AS, et al. Identification of the genes involved in odorant reception and detection in the palm weevil *Rhynchophorus ferrugineus*, an important quarantine pest, by antennal transcriptome analysis. BMC Genomics. 2016;**17**(69). DOI: 10.1186/s12864-016-2362-6

[113] Antony B, Johny J, Aldosari S, Abdelazim M. Identification and expression profiling of novel plant

cell wall degrading enzymes from a destructive pest of palm trees, *Rhynchophorus ferrugineus*. Insect Molecular Biology. 2017;**26**:469-484. DOI: 10.1111/imb.12314

[114] Antony B, Johny J, Abdelazim MM, Jakše J, Al-Saleh MA, Pain A. Global transcriptome profiling and functional analysis reveal that tissue-specific constitutive overexpression of cytochrome P450s confers tolerance to imidacloprid in palm weevils in date palm fields. BMC Genomics. 2019. DOI: 10.1186/s12864-019-5837-4

Chapter 2

Recent Trends in the Early Detection of the Invasive Red Palm Weevil, *Rhynchophorus ferrugineus* (Olivier)

Maged E.A. Mohammed, Hamadttu A.F. El-Shafie and Mohammed R. Alhajhoj

Abstract

Red palm weevil (RPW), *Rhynchophorus ferrugineus* (Olivier), is one of the most invasive pest species that poses a serious threat to date palm and coconut palm cultivation as well as the ornamental Canary Island palm. RPW causes massive economic losses in the date palm production sector worldwide. The most important challenge of RPW detection in the early stages of an infestation is the presence of a few externally visible signs. Infested palm shows visible signs when the infestation is more advanced; in this case, the rescuing of infested palms is more complicated. Early detection is a useful tool to eradicate and control RPW successfully. Until now, the early detection techniques of RPW rely mainly on visual inspection and pheromone trapping. Several methods to detect RPW infestation have recently emerged. These include remote sensing, highly sensitive microphones, thermal sensors, drones, acoustic sensors, and sniffer dogs. The main objective of this chapter is to provide an overview of the modern methods for early detection of the RPW and discuss the most important RPW detection technologies that are field applicable.

Keywords: red palm weevil, date palm, early detection, remote sensing systems, acoustic sensors, ornamental palms

1. Introduction

Smart and precision agriculture are the most evolving trends in the modern agriculture industry. The electronic sensors can be embedded into the plants to obtain necessary data for aiding in decision-making to detect pest infestation or to improve crop quality. Concerning the date palm, there is an increasing demand to apply modern technology and smart agriculture for early detection of the RPW in the date palm plantations or other palms used for landscaping. RPW is the most dangerous pest for palms worldwide, which can cause irreversible damages, particularly at the late stage of infestation. Therefore, RPW, referred to as the palm cancer, is considered one of the most dangerous pests in the Mediterranean, Gulf Cooperation Council, part of Europe, and East Asia facing date and oil palm tree [1]. The most challenging problem in the control of RPW is the early detection of infestation, which is difficult because symptoms caused by the weevil are only visible when an infestation is more advanced. The highly infested palms are often destroyed and removed [2]. For this purpose, there is a requirement to use advanced technology for the early detection of the RPW. Early detection is an effective solution to control and eradicate RPW successfully. So far, the RPW detection occurs depending on the visual inspection by laborers, which is entrusted by private or public organizations. The inspection by laborers is very costly for owners of private farms; it is likely not to be implemented due to the high cost [3]. Therefore, the frequent inspection has never been applied systematically and on a large scale, even when the inspections are compulsory [4]. Although, the early detection of the RPW presence could mean a significant economization in the capital, investment in the farms, and providing job opportunities for agricultural workers [2]. RPW early detection is a major challenge due to the cryptic nature of the weevil in most of its developmental stages. RPW larva is the most dangerous stage in the pest life due to the direct destruction it causes on the infested palms. Thus, most of the early detection methods concentrate on this stage of the RPW life cycle [2, 5], although the presence of RPW adults in the palm plantation is one of the most indicative proofs of the infestation. Governments have successfully and widely used pheromone traps as the primary protocol in the integrated management to control the RPW [6, 7]. For that, the aggregation pheromone traps of RPW were included in the integrated pest management (IPM) as an essential technique to control RPW. There were unique developments in the trap designs, color, and trap density, besides the improvement of trap catching by the addition of kairomones in various forms. However, limited human resources and high transportation costs reduced the rate of monitoring and mass trapping of RPW adults [3]. Recently, there are strong ongoing efforts to develop a reliable and quick system for early detection of RPW using a combination of computer science, sensors, and modern electronic technologies. The most important and promising technologies are X-ray [2, 8, 9], acoustic systems [5, 10–17], remote sensing systems [18–21], and radio telemetry [7]. To control the RPW, it is necessary to implement an innovative and practical early detection method leading to reduce the pest population as much as possible. This action is essential, as the early detection of infested date palm trees allows the owner of farms to sanitize or to eradicate them in the event of a severe infestation. The early detection followed immediately by the palm sanitization and eradication of infested palms' parts allows to limit or prevent the RPW spread to the neighboring plantations, thus eliminating the RPW as quickly as possible [4]. The physical properties of different developmental stages of the weevil, such as sound, thermal and chemical emissions, and images are used in the early detection technologies. The main objective of this chapter is to provide an overview and engineering information about the new trends in early detection of RPW and to discuss the basic principles of the most current and promising technologies for RPW detection.

2. Biology, ecology, and economic importance of RPW

The red palm weevil (RPW), *Rhynchophorus ferrugineus* (Olivier), is the most important invasive species in the genus *Rhynchophorus* that encompasses more than 10 species [22]. RPW is native to South and Southeast Asia, where it feeds on coconut, *Cocos nucifera*, as a key pest. During the past 30 years, the weevil invaded many countries around the world and is now found in 49 countries in Europe, Africa, Asia, and the American continent [23]. As far as the host range of RPW is concerned, 40 palm species are reported, including the date palm, *Phoenix dactylifera,* and the Canary Islands palm, *Phoenix canariensis* (EPPO, 2020). Thus, the international trade of palms for plantation and landscaping purposes had resulted in the vast spread of the weevil [24, 25]. The RPW has a complete metamorphosis development; it undergoes four stages, namely, egg, larva, pupa, and adult stages. The female may lay over 200 eggs in its lifetime. The eggs hatch in 2–6 days into larvae, which last 1–3 months undergoing as many as 8 or 16 larval instars. Fully grown larvae construct a cocoon from palm fibers and undergo pupation. The pupal stage lasts for 15–30 days, after which the adult emerges. The adult weevils can live for up to 3 months, depending on abiotic and biotic factors. Adult males can be distinguished from females by the presence of a tuff of bristles on the dorsal tip of the snout, which is not found in the female [24, 26]. RPW is a multivoltine species and has a high reproductive potential where many generations can be produced per year [26]. RPW infestation on date palms occurs on the trunk near the ground level as well as on young offshoots. Palms younger than 20 years are more susceptible to attack by RPW than older palms. Infestation on male palms occurs at the crown [24]. Symptoms of damage on date palms include oozing out of brown sticky fluid with fermented odor, drying of young offshoots, ejection of chewed palm tissues, and breakage of the trunk in advanced severe infestation (**Figure 1**).

On the other hand, infestation on Canary Island palm is exclusively at the crown level, and the symptoms of damage include the presence of holes on the fronds, asymmetrical growth of the crown, and eventual collapse of the palm in severe infestation [26]. No recent estimate of economic losses on palms in the world due to the invasive RPW is available. However, the severe killing of palms in invaded areas has been reported [24, 25, 27]. The ecological characteristics of high fecundity, high adaptation to the environment, high mobility, and the international trade of infested planting material mediated the introduction and spread of RPW. The weevil is highly invasive and more difficult to control due to the cryptic and endophagous behavior of the immature stages and the partial cryptic behavior of the adult weevils [24]. The concealment of all developmental stages inside the palms makes early detection of the weevil at an early stage of infestation extremely difficult. Currently, the integrated management strategies adopted against RPW include early detection of infested palms, pheromone mass trapping, preventive and curative chemical treatment, removal of severely infested palms, and phytosanitary measures (**Figure 2**) [24]. Phytosanitary measures, public participation in control, and awareness campaigns reduce the probabilities of RPW infestation [28]. Any successful pest invasion passes through three important stages: introduction,

Figure 1.
Visual symptoms of red palm weevil damage on a young date palm.

Figure 2.
Conventional pheromone-food-bait trap for detection and monitoring of adult red palm weevil.

establishment, and spread. Quarantine regulations, invasion monitoring, and eradication are the main management strategies to prevent the establishment and spread of RPW. Eradication of an invasive species depends largely on correct identification and early detection before the species is fully established in the ecosystem.

3. Early detection technologies

Individual visual inspection of palms in public areas or commercial plantations is unfeasible, inaccurate, and highly laborious (**Figure 3**). The direct inspection may be impossible sometimes due to palm trees being not accessible or in cases

Figure 3.
Visual inspection and removal of date palm highly infested by the red palm weevil.

where the infestations occur below the crown of palm trees. This condition requires general scale, areawide detection techniques. Below, we present recently developed detection methods and the ongoing efforts based on the promising technologies for the early detection of the RPW infestations in the palm trees. In particular, we focus here on the most promising technologies such as acoustic systems, remote sensing systems, and the X-ray.

3.1 Acoustic technology

Overview: The acoustic waves have longitudinal and transverse motion shapes in the solid and involve mechanical movements of molecules or atoms, unlike electromagnetic waves that have transverse waves motion in nature and involve the oscillation of magnetic and electric fields. Acoustic waves show numerous properties, including frequency, wavelength, amplitude, and period; through these properties, the sound can be distinguished. The acoustic waves are heard by the human ear as sound at a frequency ranging from 20 Hz to 20 kHz. They can be defined as longitudinal mechanical waves that occur by the pressure oscillation that moves through a gas, solid, or liquid in the pattern of a wave [29]. This definition is extended to the solids where the waves are described as infrasound, sound, or ultrasound based on whether the wave frequency is below, in the range, or higher than the audible limit. In widespread usage, the expression of acoustics refers to all mechanical wave types in gases, liquids, and solids. Ultrasonic is a sound wave that has a frequency higher than the human hearing range. Ultrasonic energy is produced by a longitudinal mechanical wave with a vibration frequency of over 20 kHz with one-dimensional propagation, as shown in **Figure 1**. The signal of acoustic occurs in the field from several sources such as the insect movements and chewing during its feeding, the turbulence of air and gases, flowing of fluids, the impact of solid, and plant movements due to the air. The acoustic waves are a natural phenomenon; it may include just individual frequency as a pure steady-state sine wave (**Figure 4**) or contain multiple frequencies, such as the noise generated by several sound sources [30]. An acoustic sensor is an electronic tool that produces an output signal in response to a specific sound input. The output signal of the acoustic sensor is usually digital voltage pulse stream, electrical current or analog voltage, or an oscillated voltage with frequency suited with the input quantity value. The acoustic sensors are characterized in many various methods, such as sensitivity, accuracy, and selectivity. These characteristics determine the quality of the acoustic sensor in terms of the magnitude measuring of the output signal generated in response to a delivered magnitude input and the minimum limit of change in the input signal that measures and characterizes it. The development of integrated circuits (ICs) decreased the sensor and computing costs and produced economic systems sophisticated to deal with signals from the different acoustic sensors [31]. The detection

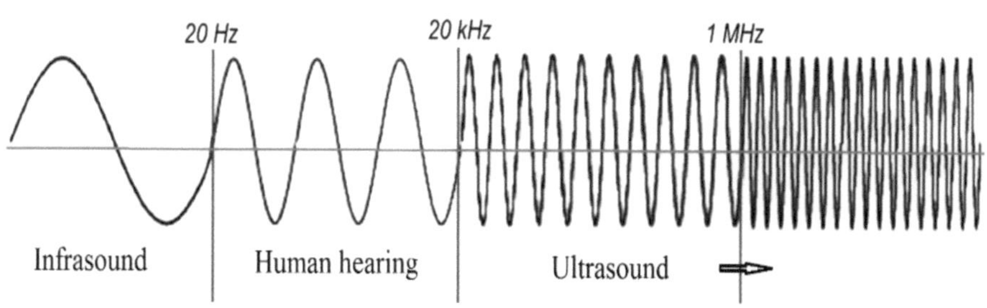

Figure 4.
Sound spectrum.

systems are based on acoustic sensors, due to the insect activities' sound inside the tree, which can be heard by humans. Moreover, under reasonable noise levels, the sound frequency of the insects can be isolated from the background noise sound and regular environment acoustic. The current models of RPW detection principally insert the acoustic probe in the palm trunk to obtain RPW sound in real time. The variations among these models are essentially in signal processing procedures, but all involved systems demand field monitoring. Field location has various background noises; the acoustic inspections must be performed in a quiet background, typically early in the day to improve detectability. With this procedure adoption, acoustic technology has been successful in detecting RPW in field inspections in many countries [3, 15].

Application of acoustic technology: Acoustic technology has the potential to be used in RPW early detection. The most applicable methods of acoustic detection for RPW depend on identifying the gnawing sounds of RPW larvae, which occurs during their movement and chewing inside the infested palm trees. The sounds of the RPW and the sound emitted from the healthy palm trees and the other sound sources are usually difficult to separate. There are many efforts to solve this problem; for example, a research was conducted to develop signal processing software devices that identify signal features and selectively magnify insect-produced sounds that distinguish sounds produced by the target insect from those produced by other sources [14]. However, the workers must recognize where to insert probes of acoustic and must use complicated signal analyses to help identify RPW sounds from other noise of background and insects. The recent decrease in the costs of microcontroller systems and increases in computing power have produced great solutions in the application of acoustic technology for RPW early detection using the acoustic probes [3]. Therefore, a mathematical method was developed to automatically detect the acoustic activity of RPW in palm offshoots and to implement it in a prototype setup. The algorithm successfully obtained detection rates as high as 98.9%. The results show that it is possible to detect the larvae of RPW sounds using the mathematical method using commercial recording devices and speech recognition [32]. Rach et al. designed a bioacoustic sensor, which was placed in the palm tree. The designed sensor was ready to analyze the captured sound signal through long periods. The sound results were connected wirelessly to the control station for stored and finally processed. The prototype was programmed to send warning messages when the sound signal reached the predefined alarm thresholds. The advantage of this prototype is to allow the supervisors to monitor the status of the palm trees orchards online with achieving detection rates over 90% [10]. A piezoelectric sensor was applied in the field to detect the sound of RPW larvae. The results were concluded that the detection using this sensor was efficient in detecting infested trees with the RPW with a sensitivity of 80% [33]. Mao et al. used an optical fiber distributed acoustic sensor as a solution to early detection of the RPW in the lab environment as a condition similar to the farms that include a palm tree infested with RPW. They recorded the sound of RPW through a loudspeaker set inside the palm trunk. For the noise sources, they used a fan to blow the air toward the palm tree and used a loudspeaker to make sounds of the birds around the palm tree. Under these laboratory conditions, they mentioned that the acoustic system could detect the sound of RPW at different positions along with the optical fiber [15]. The optical fiber distributed acoustic sensor has also been used in the Kingdom of Saudi Arabia (KSA) as an alternative technology for the early detection of RPW. The applied sensitive sensor exposes the feeding sound detection created by 12-day-old RPW larvae in the infested palm tree. In comparison with the commonly suggested technologies, this sensing system represents a noninvasive alternative, cost-effective method that could provide monitoring for 1000 palms or even more in real time, for the whole day [5]. The acoustic emissions created by the RPW inside the date palm were measured and recorded successfully. The time-frequency techniques were utilized to carry out the analysis procedure [34]. In another study, the acoustic system was used to detect the infestations of date palm trees by RPW. The system was evaluated in terms of its ability to detect the RPW in the early stage inside the tested palm tree. The technique of signal processing, known as the time-frequency analysis, was assessed in terms of the accuracy of the system in recognizing the RPW acoustic signature after the acoustic signal acquisitions were completed. The system experimented in the laboratory first then operated in the field on infested or suspected palm trees. The results showed that the acoustic monitoring procedure, besides the technique of signal processing, is very hopeful for the early detection of the RPW larva and the adult in the date palm trees [16]. A signal processing system was developed to detect the infestation of RPW based on the bioacoustic characteristics of their feeding sound. The developed system was investigated and had higher than 94% efficiency and can effectively detect the RPW feeding sound in a sound stream of 5 min. The results indicate the effectiveness of the developed system with the selected characteristics, functions of the window, and period of the frame to detect the RPW through the sound of its feeding [35]. In another study, the electronic device based on acoustic sensors was also developed for early detection of RPW larvae in the interior of the date palm trees. The developed device was based on specific frequencies rationally correlated to the larvae of RPW feeding activity. The finding concluded that it is possible to detect larvae activity (2-week old) by investigating the sound intensity of about 2250 Hz in the date palms infested (only with five individuals) under controlled environmental conditions. The other insect's activity artificially located in the interior of date palms did not affect the device's response [13]. The infested palms were detected using a developed acoustic system with accuracy over 97%. Four infested positions of palm trees (at the bases of the two lowermost leaves and at either side of the palm base) were recognized with high accuracy. The device contains a sensor to obtain the sounds of red palm weevil larvae, a set of headphones to receive the output sound by the user, and an electronic unit that processes the received sounds, and it can be installed on the palm [17]. A study was carried under the hypothesis that temporal features and characteristic spectral features in sounds of RPW larvae can be mixed to create improved indicators for automated detection of date palm infestations. For this objective, a signal processing system was developed with the available acoustic technology to detect the presence of RPW larva in the palm tree through its sounds of feeding. The features were extracted, including alternative features such as temporal slope, spread, and roll-off. The results validate the effectiveness of the developed system to detect the existence of the RPW. The economic damage due to RPW in the date palm could be decreased significantly by the bioacoustic identification in an earlier stage of RPW infestation and by applying the suitable treatment [11].

3.2 Remote sensing system

Overview: The remote sensing system has been classified into five groups (**Figure 5**) based on remote sensors. Most of these sensors have been used for plant pest detection to provide information on the processes of physiological and plant chemical parameters under stress resulting from the pest infestation. The thermal remote sensing technology or thermography is a nondestructive method used to define the thermal characteristics of any object. The principle of thermography is dependent on converting invisible radiation patterns (infrared radiation) of any object into visible thermal images [18, 20]. All objects emit infrared radiation energy when it has a temperature above absolute zero. Consequently, irrespective

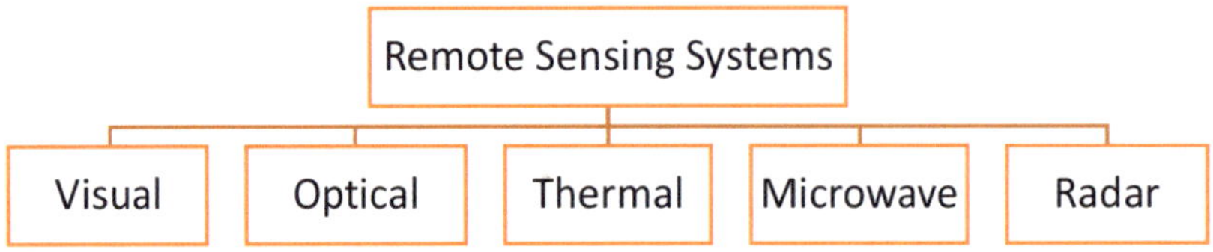

Figure 5.
Remote sensing systems [18].

of ambient light, the thermal cameras can detect all objects. The spectrum and the amount of thermal radiation depend directly on the object's surface temperature, the possibility of the thermal imaging camera to detect the object temperature. The emissivity of the object also influences radiation, which limits the accuracy of the system. On the other hand, the radiation absorption by the atmosphere and the radiation that originates from the surroundings can all affect the accuracy of the thermal imaging. In thermal remote sensing, the patterns of objects' radiation are converted into visible images named thermal images. The thermal images can be obtained by connecting portable thermal sensors with optical systems mounted on an airplane or a satellite. The thermal imaging can be classified depending on the excitation of the samples as two types, namely: active thermography and passive thermography. In the active thermography, an external source is applied to excite the test sample, thus producing temperature variation among the non-defective and defective sections within the sample. The excitation can be achieved through mechanical-, optical-, electromagnetic-, heating-, and convection-based cooling or heating. In the passive thermography, the natural emission of infrared from the object is used to detect any differences in its structure. The temperature difference among the defective objects and their surroundings is used to detect it [36].

Application of remote sensing system and thermal imaging: Remote sensing system and thermal imaging applications are the most alternative approaches for RPW detection, monitoring the physiological changes happening in the palm tree due to RPW infestations [3, 18, 37]. Farquad proposed the use of thermal remote sensing technology (thermography) for RPW early detection in order to take remedial actions at the earliest possible stage for date palm loss reduction [20]. The thermal detection is applied based on two monitoring types: the first type occurs by monitoring the local temperature in the infested part, and the second type occurs by monitoring the water stress of the palm tree. The first type depends on monitoring the local temperature in the palm tree caused by the intensive fermentation of plant tissues due to the feeding activity of the RPW larvae [38]. Unfortunately, this monitoring type is based on the change in the temperature that occurs only inside the palm or in the crown center. Due to the fermenting of the tissues in the heavy infestations. Due to the natural insulation is given by the palm's tissue, the outside surface temperature equilibrates with the ambient temperature causing difficulty in monitoring the internal temperature changes. The internal temperature changes can only be monitored via temperature probes inserted into the palm in the infested part. The second type depends on monitoring the water stress of the palm tree due to injuries in the vascular system of the palm created by larvae tunneling in the palm tissue. The decrease in stomatal conductance of the palm tree increases in the temperature of the palm canopy that can be thermally detected. The temperature changes of the palm canopy were successfully used to detect the RPW infestation in the date palms through the thermal portion inspection of the reflected irradiation spectrum before any appearance of visible detected signs that resulted from RPW infestation. The thermal images of the palm canopy can be obtained remotely by aerial photography using airplanes or drones. Generally, advanced thermal technology offers the potential to obtain spatial data to facilitate the mapping of the wide area at one aerial imaging operation for palm canopy temperatures. Thus, the palm map based on water stress distribution can be provided to recognize palm trees assumed of being infested [3, 39, 40]. Remote detection by thermal imaging based on the irradiation emitted from the thermal spectrum of the physiological changes in infested palms and palm tree canopy can produce great outcomes. The systems of thermal imaging and their application are considered within the promising technology on the RPW early detection. Therefore, the deployment of thermal image systems can obtain a good solution to RPW detection [36]. El-Faki et al. investigated thermal aspect potentials of date palm infested by RPW in early detection. The artificial infestation using males and females of RPW was done for date palm trunk. The effects of three infestations were examined during a 24-day period. The replicated measures analysis indicated that the temperatures of the healthy palms during two seasons were 31.83 and 27.56°C and the temperatures of infested palms through the two seasons were 33.22 and 30.08°C. While the corresponding ambient temperatures were 31.83 and 28.03°C, respectively. They mentioned that the study provides important information for developing a system based on a real-time sensor for nondestructive early detection of RPW [38]. Bannari et al. conducted a study on palm trees using water stress indices to discriminate among different stages of RPW stress attacks. They assessed the different water stress indices using the technology of Worldview-3 data. Based on field identification, five palm tree categories were studied: healthy trees, infested with RPW (treated), infested with RPW (untreated), severely infested, and dead. Spectral determinations were obtained for every sample using the analytical spectral devices. The results revealed that the water stress indices are sensitive to palm water agitation produced by RPW infestation. Based on their results, remote sensing using the Worldview-3 data is a promising alternative technology for RPW detection based on water stress indices [19]. Koubaa 2019 et al. developed a prototype-based smart palm tree monitoring that allows monitoring palm trees remotely using smart agriculture sensors and provides early detection of the RPW. This prototype enables the users to follow their palm farms for early detection of RPW infestations through interactions with date palm farms using mobile and web applications. They used the "Elm company IoT platform" to interface among the user layer and sensor layer. They have collected data utilizing sensors of accelerometer and applied the processing of signal and statistical methods to analyze collected data and to determine the infestation fingerprint [37]. Ghulam Rasool et al. evaluated the efficacy of some noninvasive optical devices including a thermal camera, digital camera, resisttograph, Tree Radar Unit (TRU™), magnetic, near-infrared spectroscopy (NIR), and DNA biosensor to detect RPW infestation (under field conditions) in date palm trees in the Kingdom of Saudi Arabia. They mentioned that after the date palm trees were inspected with the different devices, each tree was dissected in detail to validate each device's accuracy. The results found that the visual RPW detection approach presented the highest accuracy of 87%, followed by accuracies of 77, 73, 73, 61, and 52% for devices of Radar 2000, Radar 900, resistograph, thermal camera, and digital camera, respectively. Absorption spectra produced during near-infrared for the samples of date palm tissue that were infested, wounded, and in control showed a difference in the corresponding peak gradient between 1850 and 1950 nm. Another experiment by Ghulam Rasool et al. also determined the DNA biosensor efficiency for detecting RPW adults as 100%, followed by 83% for pupae, 63% for larvae, 60% for eggs, and 39% for control. Based on the tested device's efficiency, the near-infrared spectroscopy (NIR) and resistograph have the best potential to detect the infestation by RPW in the date palm trees [41]. Massimo et al. also tested a similar noninvasive approach (thermal camera, digital camera,

Tree Radar Unit TRU™, a densitometer, and a penetrometer) to detect RPW in Italy and KSA. In Italy, the thermal camera showed a high accuracy of 96.29% and the digital camera showed an accuracy of 92.6% compared to close visual inspection. Tree Radar Unit and densitometer additionally showed great accuracy of 83.33 and 88.9%, respectively. In the Kingdom of Saudi Arabia, the thermal camera showed an accuracy of 77.7% compared to invasive determination. While the digital camera revealed a lower accuracy of 66.7%. TRU™ also provided an accuracy of 74.7% compared to invasive diagnosis. The satisfying conclusions taken using the application of remote sensing are good as a starting point to develop an integrated protocol for the early detection of RPW and control strategies [40]. A study was conducted to analyze the fitness of "8-band WorldView-2 satellite imagery" for detecting the tree infested by bark beetle at two intensity stages of dead and green attack versus the healthy trees. The results showed that "WorldView-2 satellite" information might be helpful for large-scale applications aiming at early detection of bark beetle infestation with the remote sensing data. Despite this, the information seemed uncertain to identify each tree at the green attack of the infestation stage, but this information can be used for the remotely sensed maps of bark beetle infestation, which can be used also positively as a contributor to the bark beetle modeling community. The remotely sensed maps can be used also as an alternative application to cost-intensive record data for GIS modeling approaches [42].

3.3 X-ray

Overview: X-rays are forms of electromagnetic radiation (similar to visible lights), as shown in **Figure 6**, but unlike the visible light, X-rays have higher energy and can pass through most materials and objects including the material to be tested due to its high frequency (10^{18} Hz) and low wavelength (10^{-10} m). X-rays are used in a quick nondestructive test that produces images of the different objects and structures inside the material. The beams of X-ray pass through the material, and they are absorbed in different amounts in the materials they pass through depending on the density of each material. Whenever the material is dense, it appears as white color, while the fatty objects appear as gray shades, and the empty spaces show up as black. X-rays are used to create images of objects, hole space, and tissues in the trunk of palm trees. When X-rays pass through the material, it will also pass through a detector of the X-ray on the other side, creating an image based on the shadows, which are formed by the objects, tissues, and empty spaces inside the material. In agricultural applications, types of detectors of X-rays that produce digital images are used. The digital X-ray image produced from this case is defined as "radiograph" [43].

Application of X-rays: To detect the RPW using a radiograph, two X-rays parts (X-ray source and X-ray detector) are placed around the trunk of the palm tree (in

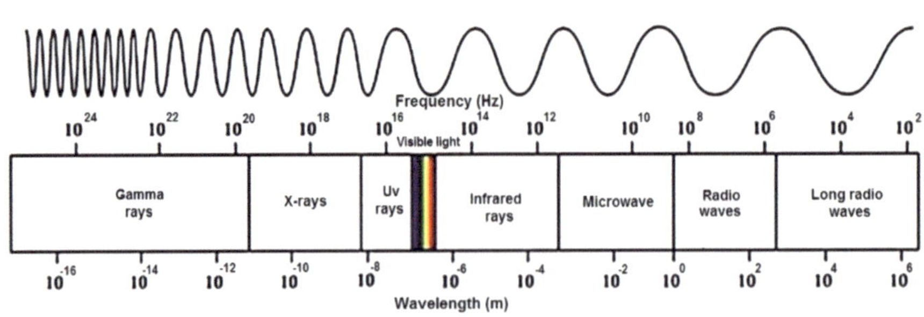

Figure 6.
The electromagnetic spectrum as a function of wavelength (m) and frequency (Hz).

a completely facing) so that the part of the tree to be imaged is located between the source and detector of X-rays. The X-rays will move through the trunk tissues, tunnels (made by insects), and the RPW body (larvae, pupae, or adult stages). Depending on the radiological density of each tissue, the X-rays pass through it; the X-ray will be absorbed in various amounts. Both the atomic number (proton's number in the nucleus of atoms) and the material density to be imaged limit the radiological density. The objects, such as the RPW, contain protein and fatty substances in all stages, which have a higher atomic number than the trunk's tissues. Therefore, the RPW's body will absorb X-rays and produce more contrast than any tissue on the X-ray detector; thus, the adult, pupae, or larvae of RPW will appear whiter than all other tissues on the radiograph detector.

On the other hand, when the X-rays move quickly through less radiologically in the material, such as the wood materials and trunk tissue, the material is displayed in shades of dark gray on the radiograph. On the contrary, when the X-rays pass through empty holes and tunnels resulted from the insects, the X-rays do not absorb it, and these holes will be displayed in black color on the radiograph. Alghamdi mentioned that using X-ray imaging systems can present direct and easy results with some modifications where the sensitivity is very low for light elements [44]. Ma et al. carried out a simulation using X-rays CT applied for small size RPW detection with the Monte Carlo method; the simulation demonstrated that the cone-beam CT reconstruction from the projections' finite number might be sufficient for the RPW detection. They determined that the technique is applicable in the farms with a simple modification of the available X-ray apparatus [2]. Haff and Slaughter constructed a high-resolution X-ray imaging system using a low-energy X-ray image intensifier, CCD camera connected with a PC, and a high-resolution real-time X-ray (low-energy and high-current X-ray source) to detect the granary weevil in the wheat kernels. The field of view was 6 cm^2, which is enough to inspect approximately 350 grains in a single frame with exposure time of 149 ms. The percentage of images correctly classified averaged 90.2% for the film observations compared to 84.4% for the real-time system [9].

4. Conclusion

Red palm weevil (RPW) causes massive economic losses in the date palm production sector as well as ornamental palms used in landscape worldwide. The most important challenges of RPW detection in the early stages are the presence of a few externally visible signs upon which detection can be based. Early detection is essential to initiate a rapid response to the eradication and effective treatment of RPW. The most effective techniques currently used for early detection of the RPW rely mainly on visual inspection and pheromone trapping. Dependence on the traditional method of visual inspection of palm trees to detect the RPW and other pest infestations is both time consuming and laborious. Additionally, visual inspection is only effective in a more advanced stage of infestation. Recently, the emphasis has been on using the modern technologies and development of systems and devices for early detection of the infestations in palm trees, especially the infestation by RPW. The most important applications of experimented technology were X-rays, remote sensing systems, acoustic sensors and software, bioacoustic applications, thermal image, X-rays, CT, etc. Modern alternative techniques have been undertaken for the early detection of the RPW infestations to reduce the dependence on visual inspection, which is expensive and requires highly trained labor. Most modern detection methods have their strength and weaknesses. However, an effective, applicable, and direct method for early detection of RPW, at all life stages inside the palm trees, remains to be developed. Therefore, to increase the overall reliability and efficiency of early detection, there is a need for further research and experiments to develop and produce reliable, easy-to-handle, and cost-effective portable detection systems for RPW early detection. Based on the findings of previous studies, we particularly mention promising technologies such as acoustic and remote sensing systems, thermal imaging, and the X-rays.

Acknowledgements

The authors gratefully acknowledge the financial support for the project number DPRC-6-(2018) by the Date Palm Research Center of Excellence, King Faisal University, Kingdom of Saudi Arabia.

Conflict of interest

The authors declare no conflict of interest.

Author details

Maged E.A. Mohammed[1,2]*, Hamadttu A.F. El-Shafie[1] and Mohammed R. Alhajhoj[1]

1 Date Palm Research Center of Excellence, King Faisal University, KSA

2 Agricultural Engineering Department, Faculty of Agriculture, Menoufia University, Egypt

*Address all correspondence to: maged3929@yahoo.ca

References

[1] Tofailli K. The early detection of red palm weevil: A new method. In: IV International Date Palm Conference; 2010. pp. 441-449. DOI: 10.17660/ActaHortic.2010.882.50

[2] Ma AKW, Alghamdi AA, Tofailli K, Spyrou NM. X-ray CT in the detection of palm weevils. Journal of Radioanalytical and Nuclear Chemistry. 2012;**291**(2):353-357. DOI: 10.1007/s10967-011-1202-z

[3] Food and Agriculture Organization of the United (FAO) & International Center for Advanced Mediterranean Agronomic Studies (CIHEAM). The Scientific Consultation and High-Level Meeting on Red Palm Weevil Management; 2017. pp. 29-31. Available from: http://www.fao.org/3/a-bu018e.pdf

[4] Ferry M, Cousin R, Chabernaud D, Ferrero F. An effective strategy to obtain very rapidly the red palm weevil decline in an area planted with ornamental palms. Arab Journal of Plant Protection. 2019;**37**(2):188-197. DOI: 10.22268/AJPP-037.2.188197

[5] Ashry I, Mao Y, Al-Fehaid Y, Al-Shawaf A, Al-Bagshi M, Al-Brahim S, et al. Early detection of red palm weevil using distributed optical sensor. Scientific Reports. 2020;**10**(1):3155. DOI: 10.1038/s41598-020-60171-7

[6] Ellis SE. New Pest Response Guidelines Spodoptera. United States Department of Agriculture Animal and Plant Health Inspection Service; 2004. Available from: http://www.aphis.usda.gov/ppq/manuals/

[7] Al Ansi A, Aldryhim Y, Al JA. First use of radio telemetry to assess behavior of red palm weevil, *Rhynchophorus ferrugineus* (Olivier) (Coleoptera: Dryophthoridae) in the presence and absence of pheromone traps. Computers and Electronics in Agriculture. 2020;**170**:105252. DOI: 10.1016/j.compag.2020.105252

[8] Krähenbühl A, Kerautret B, Debled-Rennesson I, Longuetaud F, Mothe F. Knot detection in X-ray CT images of wood. In: Lecture Notes in Computer Science (Including Subseries Lecture Notes in Artificial Intelligence and Lecture Notes in Bioinformatics). 2012. pp. 209-218. DOI: 10.1007/978-3-642-33191-6_21

[9] Haff RP, Slaughter DC. Real-time x-ray inspection of wheat for infestation by the granary weevil, *Sitophilus granarius* (L.). Transactions of the American Society of Agricultural Engineers. 2004;**47**(2):531-537. DOI: 10.13031/2013.16022

[10] Rach MM, Gomis HM, Granado OL, Malumbres MP, Campoy AM, Martín JJS. On the design of a bioacoustic sensor for the early detection of the red palm weevil. Sensors (Switzerland). 2013;**13**(2):1706-1729. DOI: 10.3390/s130201706

[11] Hussein WB, Hussein MA, Becker T. Detection of the red palm weevil rhynchophorus ferrugineus using its bioacoustics features. Bioacoustics. 2010;**19**(3):177-194

[12] Gong CSA, Lee HC, Chuang YC, Li TH, Su CHS, Huang LH, et al. Design and implementation of acoustic sensing system for online early fault detection in industrial fans. Journal of Sensors. 2018;**2018**. DOI: 10.1155/2018/4105208

[13] Gutiérrez A, Ruiz V, Moltó E, Tapia G, del Mar Téllez M. Development of a bioacoustic sensor for the early detection of red palm weevil (*Rhynchophorus ferrugineus* Olivier). Crop Protection. 2010;**29**(7):671-676. DOI: 10.1016/j.cropro.2010.02.001

[14] Mankin RW. Recent developments in the use of acoustic sensors and

signal processing tools to target early infestations of red palm weevil in agricultural environments. Florida Entomologist. 2011;**94**(4):761-765. DOI: 10.1653/024.094.0405

[15] Mao Y, Ashry I, Ng TK, Ooi BS. Towards early detection of red palm weevil using optical fiber distributed acoustic sensor. In: 2019 Opt Fiber Commun Conf Exhib OSA Technical Digest Optical Society of America - Proc. Vol. 1. 2019. pp. 1-3. DOI: 10.1364/ofc.2019.w2a.15

[16] Al-Manie MA, Alkanhal MI. Acoustic detection of the red date palm weevil. Proceedings of World Academy of Science, Engineering and Technology. 2005;**2**(2):157-160

[17] Siriwardena KAP, Fernando LCP, Nanayakkara N, Perera KFG, Kumara ADNT, Nanayakkara T. Portable acoustic device for detection of coconut palms infested by *Rynchophorus ferrugineus* (Coleoptera: Curculionidae). Crop Protection. 2010;**29**(1):25-29. DOI: 10.1016/j.cropro.2009.09.002

[18] Al-doski J, Shattri BM, Helmi-Zulhai BM-S. Thermal imaging for pests detecting—A review. nternational Journal of Agriculture for Plant. 2016;**2**:10-30

[19] Bannari A, Mohamed AMA, El-Battay A. Water stress detection as an indicator of red palm weevil attack using worldview-3 data. International Geoscience and Remote Sensing Symposium. 2017;**2017**:4000-4003. DOI: 10.1109/IGARSS.2017.8127877

[20] Farquad MAH. Machine learning based early detection of red palm weevil using remote sensing technology in Saudi Arabia. 2018;**3**(5):587-595

[21] Al-Mashreki MH, Bin MAJ, Abd Rahim S, Md Desa K, Ali Rahman Z. Remote sensing and GIS application for assessment of land suitability potential for agriculture in the IBB governorate, the Republic of Yemen. Pakistan Journal of Biological Sciences. 2010;**13**(23):1116-1128. DOI: 10.3923/pjbs.2010.1116.1128

[22] Wattanapongsiri A. A revision of the genera *Rhynchophorus* and *Dynamis* (Coleoptera: Curculionidae). Department of Agriculture, Science Bulletin. 1966;**1**(1):418. DOI: 1957/10673

[23] EPPO. *Rhynchophorus ferrugineus*. EPPO Datasheets on Pests Recommended for Regulation. Vol. 45. 2020. Available from: https://gd.eppo.int

[24] Faleiro JR. A review of the issues and management of the red palm weevil *Rhynchophorus ferrugineus* (Coleoptera: Rhynchophoridae) in coconut and date palm during the last one hundred years. International Journal of Tropical Insect Science. 2006;**26**(3):135-154. DOI: 10.1079/IJT2006113

[25] Milosavljević I, HAF E-S, Faleiro JR, Hoddle CD, Lewis M, Hoddle MS. Palmageddon: The wasting of ornamental palms by invasive palm weevils, *Rhynchophorus* spp. Journal of Pest Science. 2004). 2019 Jan 20;**92**(1):143-156. DOI: 10.1007/s10340-018-1044-3

[26] Dembilio Ó, Jacas JA. Basic bio-ecological parameters of the invasive red palm weevil, *Rhynchophorus ferrugineus* (Coleoptera: Curculionidae), in phoenix canariensis under Mediterranean climate. Bulletin of Entomological Research. 2011 Apr 8;**101**(2):153-163. DOI: 10.1017/S0007485310000283

[27] El-Sabea AMR, Faleiro JR, Abo-El-saad MM. The threat of red palm weevil *Rhynchophorus ferrugineus* to date plantations of the Gulf region in the middle-east: An economic perspective. Outlooks on Pest Management. 2009;**20**(3):131-134. DOI: 10.1564/20jun11

[28] Sanz-Aguilar A, Cortés I, Gascón I, Martínez O, Ginard S, Tavecchia G. Modelling pest dynamics under uncertainty in pest detection: The case of the red palm weevil. Biological Invasions. 2020;**22**(5):1635-1645. DOI: 10.1007/s10530-020-02208-6

[29] Leisure RG, Leisure RG. Acoustic waves in solids. In: Ultrasonic Spectroscopy. Cambridge: Cambridge University Press; 2017. p. 56-93. DOI: 10.1017/9781316658901.004 [cited 27 May 2020]

[30] Mohammed MEA, Alhajhoj MR. Importance and applications of ultrasonic technology to improve food quality. In: Food Processing. IntechOpen; 2020. DOI: 10.5772/intechopen.88523

[31] Ballantine Jr DS, White RM, Martin SJ, Ricco AJ, Zellers ET, Frye GC, et al. Acoustic Wave Sensors: Theory, Design and Physico-Chemical Applications. Elsevier; 1996

[32] Pinhas J, Soroker V, Hetzroni A, Mizrach A, Teicher M, Goldberger J. Automatic acoustic detection of the red palm weevil. Computers and Electronics in Agriculture. 2008;**63**(2):131-139. DOI: 10.1016/j.compag.2008.02.004

[33] Hetzroni A, Soroker V, Cohen Y. Toward practical acoustic red palm weevil detection. Computers and Electronics in Agriculture. 2016;**124**:100-106. DOI: 10.1016/j.compag.2016.03.018

[34] Al-Manie MA, Alkanhal MI. Acoustic detection of the red date palm weevil. International Journal of Signal Processing. 2004;**1**:1-12

[35] Hussein WB, Hussein MA, Becker T. Application of the signal processing technology in the detection of red palm weevil. In: European Signal Processing Conference. Eusipco; 2009. pp. 1597-1601

[36] Ahmed A, Ibrahim A, Hussein S. Detection of palm tree pests using thermal imaging: A review. In: Studies in Computational Intelligence. Springer; 2019. pp. 253-270. DOI: 10.1007/978-3-030-02357-7_12

[37] Koubaa A, Aldawood A, Saeed B, Hadid A, Ahmed M, Saad A, et al. Smart Palm: An IoT Framework for Red Palm Weevil Early Detection. 2019; Available from: http://arxiv.org/abs/1910.00653

[38] El-Faki MS, El-Shafie HAF, Al-Hajhoj MBR. Potentials for early detection of red palm weevil (Coleoptera: Curculionidae)-infested date palm (Arecaceae) using temperature differentials. Canadian Entomologist. 2015;**148**(2):239-245. DOI: 10.4039/tce.2015.51

[39] Stafford JV. Precision Agriculture '15. Precision Agriculture; 2015 (April 2016). pp. 1-751. DOI: 10.3920/978-90-8686-814-8

[40] Massimo P, Alberto RA, Roberto M, Khalid AR, Ali AM. Devices to detect red palm weevil infestation on palm species. Precision Agriculture. 2018;**19**(6):1049-1061. DOI: 10.1007/s11119-018-9573-6

[41] Ghulam Rasool K, Husain M, Salman S, Tufail M, Sukirno S, Mehmood K, et al. Evaluation of some non-invasive approaches for the detection of red palm weevil infestation. Saudi Journal of Biological Sciences. 2020;**27**(1):401-406. DOI: 10.1016/j.sjbs.2019.10.010

[42] Immitzer M, Atzberger C. Early detection of bark beetle infestation in Norway spruce (*Picea abies* L.) using worldview-2 data. Photogrammetrie, Fernerkundung, Geoinformation. 2014;**2014**(5):351-367. DOI: 10.1127/1432-8364/2014/0229

[43] Ehlers J. X-ray and Neuron Reflectivity: Principles and

Applications; 1999. DOI: 10.1007/3-540-48696-8

[44] Alghamdi AA. Monte Carlo simulation of neutron tomography for palm weevil detection. Journal of Radioanalytical and Nuclear Chemistry. 2012;**291**(2):359-364

Chapter 3

Invasive Insects in India

Swoyam Singh, Jai Hind Sharma, Avinash Udikeri and Hashib Ansari

Abstract

Invasive alien species (IAS) is an exotic species that becomes established in natural or seminatural ecosystems or habitats. It is an agent of change and threatens native biological diversity. Invasive insects in India have been a major threat to household commodities, human health, agricultural produce and environment. India is highly diversified in its weather and climate, which supports the establishment of various introduced insects from other parts of the world. Furthermore, globalisation has facilitated numerous introductions of invasive insect pests. There have been 23 invasive alien insects reported from India to date. This chapter exclusively deals with the characteristics of insects that make them invasive, the common pathways of entry of the invasive insects, the steps through which the introduced insect has established itself in the foreign land and finally the ways to manage them.

Keywords: India, invasive, insects, biological control

1. Introduction

India is the seventh largest country in the world with a total area expanse of 3.2 million square kilometres (appox.) [1]. It measures 3214 km from north to south and 2933 km from east to west. India is the largest peninsula, whose three parts are covered with water, *viz.* the Arabian Sea in the west, the Bay of Bengal in the east and the Indian Ocean at the south. India, being such a huge landscape, bears various climatic conditions, soils and vegetations all around the country. This makes India one of the richest diversities in the world. The country has 15 agro-climatic zones comprising of mountains, plateaus, deserts, river, lakes, oceans and grasslands [2]. Apart from that 21.54% of the total area is covered by dense forests [3]. Hence, the complex topography, diversified climate and vegetation make India one of the favourite venues for alien species invasion. Furthermore, India being one of the fastest-growing economies has contributed an export of $330.07 billion in 2014 to as many as 190 countries in the world. On the other hand, the import statistics reveal an importation of $462.9 billion from around 140 countries around the world. This excessive trade among nations has made India an opportunistic target for the entry of the alien species [4].

Biological invasion can be well regarded as a biological pollution which causes maximum losses to the biodiversity [5]. Invasive species are threats to agricultural biodiversity as well as human and animal health. These species are non-native or exotic species which have great power of dispersal and adaptation. They are introduced unintentionally into a new area where they get a favourable climate increase in number and establish. Furthermore, the new area will be devoid of its natural enemy which unleashes the invasive species' growth without any limitation. According to the International Union for Conservation of Nature (IUCN), an invasive alien species (IAS) is an exotic species which becomes established in natural or seminatural ecosystems or habitats, is an agent of change and threatens native biological diversity. India has harboured a total of 173 invasive species including 47 invasive species of agricultural ecosystem, out of which 23 are insects [6].

2. Pathways of invasion

Pathways are the predicted routes helping the invasive species in transit to new environments. There is a long history regarding the classification of these pathways. The most common pathways include the sea, land or air. There have been several examples of entry of various species through these pathways into India. For example, cashew which is considered as one of the most important cash crops of India was introduced in the sixteenth century by the early Portuguese purposefully realising its importance, through their sea voyages [7]. Today, due to globalisation, the frequency of invasion and its consequences has increased exponentially. Some of the species were knowingly transported to a new ecosystem, while some are a matter of ignorance.

i. **Introduced as contaminants.** The trade of logs is considered to be one of the most profitable businesses. It provides a huge foreign exchange, but it comes with the consequence of introduction of alien species also. Being stored in godowns for a long period without any treatment makes these logs store house of many insects and pathogens. The importation of these logs also transmits their residing alien species [8].

There has been an importation of various food products and seeds to India from foreign countries. In the past when there was a lenient legal restriction for importation of food products, an admixture of contaminants in the food lot have been proven to be detrimental to the importing country. For example, the importation of food grains from the USA under the US PL 480 scheme (a food assistance programme of the US government to India) also brought *Parthenium hysterophorus* along with it [9].

ii. **Living industry pathway.** Living industry implies the trade of living organisms including landscaping or horticultural crops, agriculture, aquaculture/aquarium, pet trade and live seafood trade. There is every possibility that these living entities escort their respective pests while being exported. For example, Apple woolly aphids have been predicted to enter India through the importation of apple rootstock from England [10].

iii. **Transportation related.** India being rich in its cultural heritage attracts numerous tourists every year. The foreign tourist arrival rate has increased from 2.54 million per annum in 2001 to 10.56 million per annum in 2018 [11]. The increase in number of tourists has increased the chances of invasion by the following ways:

- Accidental introduction of insects or other organisms [12]
- Unintentional transit of alien species from one place to another
- Cargo shipments by air, land and water [13, 14]
- Food trade [15, 16]
- Infested fruits or vegetables carried by tourists [17]

There has been enhanced domestic conveyance with enriched networks of railways, highways, airways and harbours which further facilitate the easy spread of the introduced insects all over the country mostly by unintentional transit along with passengers.

3. Characteristics of an invasive species

- Very resilient
- Short life cycle
- Broad host range
- High dispersal ability
- Ability to withstand many environmental conditions
- High fecundity
- Voracious feeders
- Benefits from mutualist interaction

4. Steps in invasion

The process of invasion of an alien species follows certain sequential steps, *viz.* introduction, establishment, spread and naturalisation [18].

i. **Introduction.** In order to become a habitant of a new locality, beyond the natural ranges, the insect must have to first move or get itself moved from its current habitat. This movement of the insect is called as passive transport which is brought about by vectors. The most common vectors nowadays are humans or human consignments. Passive transports of these invasive species are very common and are difficult to control. Even after strong quarantine inspections, it is difficult to detect the cryptic early stages of the insect which might be transported through the tourists. The transportation of a single gravid female insect also has the ability to create havoc in the introduced environment since it is a prolific breeder and lacks its natural enemy. One of the most common examples of passive transportation of insect is the occurrence of timber beetles (Cerambycidae) in Antarctica which is predicted to be introduced through transported logs to one of the base camps [19].

Sometimes insects themselves have an inherent capacity of migration to long distances. Hence any deviations from the favourable condition make them migrate from that place. Fall armyworm, *Spodoptera frugiperda*, is one of the recent invasive insect pests in India. They *are* excellent fliers [20, 21] and can fly almost 100 kilometres in certain hours; hence they are believed to have fled from Africa to the Indian subcontinent, but still the mode of arrival of the invasive pest is uncertain. Another example is the Monarch butterfly, *Danaus plexippus* (Nymphalidae), native to North America which has been found migrating to Australia in the early nineteenth century [22].

ii. **Establishment.** Short colonisation is very common, but the insect cannot be regarded as an invasive one unless it has established itself in the new environment which is possible only when the invaded insect overcomes the environmental barriers. Unlike local colonisation, establishment of an invasive species is dependent on the amount of propagule introduced. It is believed that the establishment is promoted by disturbances conducted at the receiving environment. Disturbance promotes establishment either because native species are poorly adapted to the frequency, intensity or timing of human-mediated disturbances or because IAS are often adapted to disturbance and thus predisposed for colonisation in such environments [23]. Insects are more prone to invasiveness due to increased resource availability and decreased biotic resistance.

Global warming is another cause of rapid invasiveness of the insects. Global warming has modified the resource availability and habitat suitability, thereby deteriorating biological regime of the native insects, hence favouring the establishment of the alien insects [24]. However some of the regions might be least affected by climate change as far as the invasiveness is concerned, whereas climate change always affects negatively the native species [25].

iii. **Spread.** Spreading is the process where the initially established species spreads to other areas. Spreading is guided by environmental factors such as weather conditions, microclimate and habitat quality [26]. The community context of spreading includes interspecific competition and predators also. The decision for spreading in an individual insect is also brought about either by its behavioural responses (aggressiveness or sociality of the insect) or morphological responses (wing development as in the case of aphids). Apart from that the established insect can also spread through human-mediated transportation. The latter conserves energy for the invasive insect since the human-mediated transports are quick but the success rate of the spread is highly dependent on the habitat permeability of the habitat to which the lien insect spread [27].

5. Current status of invasive insects in India

There are 173 invasive species introduced to India including 54 terrestrial plants, 56 aquatic organisms, 47 organisms having agricultural importance and 14 organisms of island ecosystem. On the other hand, there are 23 insect pests introduced to India (**Table 1**) [6].

SI. No	Common name	Scientific name/ family	Introduced from/year	Hosts	Symptoms	Natural enemies
	Order: Hemiptera					
1.	Woolly aphid [28]	*Eriosoma lanigerum* (Aphididae)	China (1889)	Apple, pear	Nymphs and adults are sap suckers, roots are damaged by the formation of swellings, and plants look unhealthy even leading to death	**Parasitoids** *Aphelinus mali* **Predators** *Coccinella septempunctata*, *Coccinella transversoguttata*, *Balli anacharis*, *Chrysopa nigricornis*, *Exochomus uropygialis*, *Coleophora sunzetti*
2.	San Jose scale [29]	*Quadraspidiotus perniciosus* (Diaspididae)	China (1911)	*Populus* spp.; *Salix* spp.; *Aesculus* spp.; *Alnus* spp.; *Betula* spp.; *Celtis* spp.; *Fagus* spp.; *Morus* spp.	Nymphs and adults suck the sap usually from the twigs and branches even fruits and leaves depending on severity. Heavy infestation on bark causes gumming which will be fatal to plant. Pink discolouration around the parts infested by scales	**Parasitoids** *Aphytis* sp., *Novaproclia* sp., *Encarsia perniciosi* *Teleterbratus perversus* **Predator** *Chilocorus infernalis*
3.	Lantana bug [30]	*Orthezia insignis* (Orthezidae)	Sri Lanka/ West indies (1915)	*Lantana*, coffee, *Jacaranda*, *Citrus*, sweet potato, gumwood, brinjal, rose	Reproduction is parthenogenetic. Bug feeds by sucking the sap from the phloem. As a by-product, honeydew is excreted which leads to attraction of ants and development of sooty moulds (Green, 1922) resulting in foul odour and decrease of market value of fruits	**Predator** *Hyperaspis pantherina*
4.	Cottony cushion scale [29]	*Icerya purchasi* (Margarodidae)	1921	*Acacia decurrens*, *Acacia dealbata,* and also a wide range of forest trees and agriculture are affected by this insect	Insect is a sap sucker. Abundant amount of honeydew is excreted leading to the sooty mould development eventually affecting photosynthesis and yield	**Predator** *Rodolia cardinalis*
5.	Pine woolly aphid [29]	*Pineus pini* (Adelgidae)	1970	*Pinus* spp. *Pinus patula*	Aphid feeds on shoots of plants causing dieback from the tip	**Predator** *Cheilomenes aurora*

SI. No	Common name	Scientific name/ family	Introduced from/year	Hosts	Symptoms	Natural enemies
6.	Subabul psyllid [31]	*Heteropsylla cubana* (Psyllidae)	Central America (1988)	*Leucaena* sp.	These insects are sap suckers from the young shoots, leaves and inflorescences leading to complete deformation of young shoots	**Predator** *Curinus coeruleus*
7.	Spiralling whitefly [32]	*Aleurodicus dispersus* (Aleyrodidae)	Caribbean region, Central America (1993)	Wide range of plants (481 hosts)	Nymphs and adults suck sap from host plants. Premature leaf drop. Nymphs secrete white, waxy and flocculent material. Honeydew is produced which develops sooty mould affecting photosynthesis	**Parasitoids** *Encarsia haitiensis Encarsia guadeloupe* **Predators** *Axinoscymnus puttarudiah* *Oecophylla smaragdina* *Solenopsis geminate*
8.	Silver leaf whitefly [29]	*Bemisia argentifolii* (Aleyrodidae)	1999	Tomato, Squash, Poinsettia, Cucumber, Eggplants, Okra, Beans, and Cotton	Four types of damages can be seen Disorder symptoms (irregular ripening), sooty mould development due to honeydew secretion, viral disease (tomato leaf curl virus transmitted by adults) and cosmetic damage due to the presence of adults and nymphs on leaves	**Parasitoids** *Encarsia formosa*, *Eretmocerus eremicus* **Fungal pathogens** *Lecanicillium lecanii*, *Beauveria bassiana*, *Paecilomyces fumosoroseus*
9.	Papaya mealy bug [33]	*Paracoccus marginatus* (Pseudococcidae)	Central America (2007)	Mulberry, tapioca, *Jatropha*, cotton and several fruits, flowers and plantation crops	Stylets are inserted into leaves and skin of the fruits and suck the sap. Injection of toxic substance leads to chlorosis, stunting, distortion, early leaf and fruit fall. Sooty mould is developed disturbing photosynthesis	**Parasitoid** *Acerophagus papayae* **Predator** *Cryptolaemus montrouzieri*
10.	Cotton mealy bug [34]	*Phenacoccus solenopsis* (Pseudococcidae)	USA (2005)	Cotton, brinjal, okra, tomato, sesame, sunflower, rose	Sap suckers Infestation of this mealybug on cotton causes: stunting, yellowing, distortion and premature drop of leaves and fruits	**Parasitoid** *Aenasius bambawalei*

SI. No	Common name	Scientific name/ family	Introduced from/year	Hosts	Symptoms	Natural enemies
11.	Solenopsis mealy bug [35]	*Phenacoccus solenopsis* (Pseudococcidae)		Malvaceae (okra), Solanaceae (tomato, brinjal, potato, chilly), Leguminosae (field bean), Cucurbitaceae (pointed gourd, cucumber, melons and gourds)	Apart from sap sucking, they develop black sooty moulds which disturb photosynthetic activity	**Parasitoid** *Aenasius bambawalei*
12.	Rugose spiralling whitefly [36]	*Aleurodicus rugioperculatus* (Aleyrodidae)	2016	Coconut, guava, banana, mango, drumstick, jackfruit	Young ones and adults suck the sap from leaves by remaining ventral surface of the leaves	**Parasitoid** *Encarsia* sp.
13.	Woolly whitefly [37]	*Aleurothrixus floccosus* (Aleyrodidae)	Neotropical (2019)	Guava, *Citrus* species		**Parasitoid** *Cales noacki*
14.	Neotropical whitefly [38]	*Aleurotrachelus atratus* (Aleyrodidae)	Neotropical (2019)	*Cocos nucifera* and *Dypsis lutescens*		**Parasitoid** *Encarsia* spp. **Predators** *Dichochrysa astour* *Cybocephalus* spp. *Chilocorus nigrita and Jauravia pallidula*
	Order: Lepidoptera					
15.	Potato tuber moth [29]	*Phthorimaea operculella* (Gelechiidae)	Italy (1937)	Tobacco, tomato, brinjal, beet and stored potato	Caterpillars mine the leaves and bore holes on tender shoots and tubers. They can move through veins into the petiole	**Parasitoids** *Chelonus blackburni*, *Copidosoma kochleti* **Predators** *Chrysoperla zastrowi*, *Orius albidipennis*, *Labidura riparia*

SI. No	Common name	Scientific name/ family	Introduced from/year	Hosts	Symptoms	Natural enemies
16.	Diamondback moth [39]	*Plutella xylostella* (Plutellidae)	1914	Cabbage, cauliflower, radish, Knol-khol (rabi), turnip, beetroot, mustard	Mining and skeletonisation of leaves and caterpillar also bores into heads of cabbage	**Parasitoids** *Brachymeria excarinata*, *Tetrastichus sokolowskii* **Predators** *Motacilla flava*, *Tapinoma melanocephalum*, *Pheidole* spp., *Camponotus sericeus*
17.	South American tomato leaf miner [40]	*Tuta absoluta* (Gelechiidae)	South America (2014)	Tomato, potato, pepper, brinjal	Larvae attack leaves, buds, stalks and fruits. Feeding results in blotches visible from both sides, galleries on leaves and pinholes on fruits from stalk end stuffed with frass	**Parasitoid** *Trichogramma achaea* **Predator** *Nesidiocoris tenuis*
18.	Fall armyworm [41]	*Spodoptera frugiperda* (Noctuidae)	America to Africa, Africa to India (2018)	Maize, millet, sorghum, sugarcane, rice, wheat, cowpea, groundnut, potato, soybean, cotton	Eggs are laid in the whorl (inner side) and also under leaves in mass. Hatched larvae feed on leaves by scraping and skeletonising leaving a silvery transparent membrane. Feeds unopened leaves in whorls and stuffs with frass. Older larvae feed on primordial shoot and tassel leading to dead heart symptom	**Parasitoids** *Telenomus* sp., *Trichogramma* sp.
	Order: Hymenoptera					
19.	Blue gum chalcid [29]	*Leptocybe invasa* (Eulophidae)	Australia	*Eucalyptus* sp.	*L. invasa* lay eggs in the bark of shoots or the midribs of leaves. Small, white maggots produce galls on the midrib of leaves, petioles and twigs leading to gnarled appearance, stunted growth, lodging, dieback and eventually death of the tree	

Sl. No	Common name	Scientific name/ family	Introduced from/year	Hosts	Symptoms	Natural enemies
20.	Erythrina gall wasp [42]	*Quadrastichus erythrinae* (Eulophidae)	2005	*Erythrina* sp., black pepper vanilla	Wasp forms galls on the leaves, stems, petioles, and youngs. Due to galls there reduction in size and number eventually decreases the growth. One wasp per gall is present, whereas galls on shoots and petiole have five individuals. Trees affected become scrawny with crinkle leaves later leading to defoliation and death	
	Order: Coleoptera					
21.	Coffee berry borer [43]	*Hypothenemus hampei* (Scolytidae)	Northeast Africa (1990)	Arabica and robusta types of coffee	The fertilised female bores an entrance hole at the terminal pore or in the calyx ridge of the differential tissue that surrounds the pore and lays bean shaped eggs	**Parasitoids** *Prorops nasuta, Cephalonomia stephanoderis, Phymastichus coffea, Cephalonomia stephanoderis*
	Order: Diptera					
22.	Serpentine leaf miner [44]	*Liriomyza trifolii* (Agromyzidae)	USA (1990)	Pea, cucurbits, tomato, castor, ornamental plants (feeds on more than 78 annual plant species)	The adult female makes punctures in the leaf tissue with its ovipositor for both feeding and oviposition. The larvae that hatch out from the eggs mine the leaf feeding on the mesophyll region leaving a serpentine structure	**Parasitoid** *Hemiptarsenus varicornis*
	Class: Arachnida					
23.	The coconut eriophyid mite [29]	*Aceria guerreronis* (Eriophyidae)	1997	Coconut	Nuts are discoloured resulting in the reduction of market value	**Predator** *Neoseiulus baraki* **Fungal pathogen** *Hirsutella thompsonii*

Table 1.
Current status of invasive insects in India.

6. Management of invasive insects

The process of management of invasive insects (**Figure 1**) includes management at three different levels of invasion of pest:

a. When the pest has not been introduced: Preventive measures are taken to avoid the entry of the invasive insect, *viz.* pest risk analysis (PRA), quarantine and monitoring. This is the best way in managing the invasive species.

b. When the species is introduced but is not spread to nearby areas: Postquarantine measures are taken in such cases such as rejection of the consignment from which the pest has introduced and eradication by means of fumigation of the consignment lot.

c. When the introduced insect has established itself: Various curative measures such as cultural, biological and chemical means of management are adopted.

The Indian government has framed certain laws to cope with the invasive species. The formulation of laws to prevent the entry of invasive species ages back to 1914 when the Destructive Insects and Pests Act 1914 was framed [45]. With the elapse of time, these laws were modified from time to time. Presently, the primary plant quarantine concerns of India are dealt by the Plant Quarantine Order 2003. The Plant Quarantine Order 2003 includes new import policies with required statutory measures which aim to restrict the import of infested plants or plant products. The order advocates a prior PRA to estimate the phytosanitary measures required to protect plant resources against the invasive pest [46].

Plant quarantine facilities include:

- An integrated information management system
- An integrated pest risk analysis system and a national pest risk analysis unit for conducting an integrated pest surveillance
- An integrated phytosanitary border control system
- A national phytosanitary database
- A national management centre for phytosanitary certification to continuously review the national standards for export phytosanitary certification

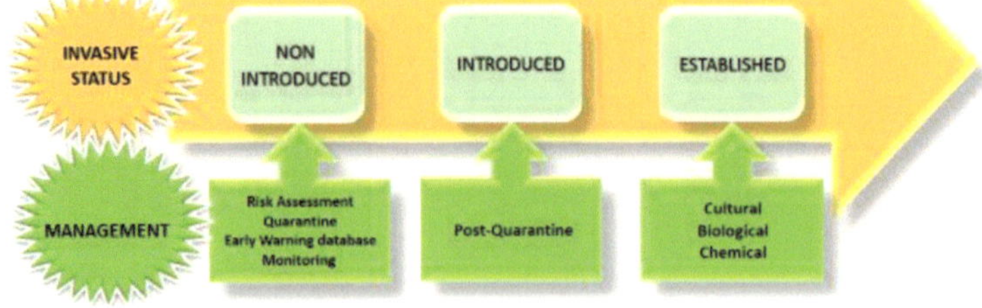

Figure 1.
Management of invasive insects.

i. **Monitoring.** There is a total of 71 plant quarantine stations across major and minor ports (34 seaports, 12 airports, 14 land frontiers and 11 foreign post offices) in India [47] which deals exclusively on restricting the import of any foreign contaminants. One of the major approaches for managing the invasive insects is its early detection. But the problem lies with the exact identification of the insect (to species level). Identification of the species requires expertise in insect taxonomy which is a limiting constraint in India [48]. Hence the government of India has established molecular diagnostic facilities across the quarantine station for easy and rapid detection of invasive insects [49]. Apart from molecular techniques, monitoring is also done by the use of chemical pheromones (sex pheromones and attractants) or by physical with the help of yellow sticky traps, light traps, etc.

ii. **Biological.** The boom reproduction of invasive insects in a new environment is because of non-availability of their natural enemy and unlimited food supply [50]. Biological control is an ancient practice to control introduced pests, which deals with a timely introduction (classical biological control), augmentation (mass release of native or exotic natural enemies) and conservation (habitat management) of natural enemy (predators and parasitoids) from their (invasive insect's) native places in hope that they may reduce the invasive pest population to non-harming levels [51]. There are many examples of biological control which are listed above (**Table 1**). Biological control is sufficient to control the alarming invasiveness if once the natural enemy is established, has long-term effects and is cost-effective too [52]. Sometimes the biological control comes with a consequence which should be pre-analysed before the introduction of the bioagent [53]. For example, *Zygogramma bicolorata* was introduced in India to manage *Parthenium hysterophorus* which ended up as a pest of sunflower [54]. There must be prior research to prevent such introduction of natural agents which may have a negative impact.

iii. **Chemical control.** Prevention is always better than cure. Hence, strict quarantine is the best solution for the management of invasive insects, that is, a thorough investigation of all kinds of imported goods and products in order to hamper the introduction of dangerous species. However, after the breaching for this barrier, the next prompt control measure is pesticides (chemicals). Pesticides are quick acting and are very efficient in reducing or eradicating the invasive insects. But prior knowledge about the insecticide regarding its mode of action, selectivity and residual effect is very much essential while applying an insecticide to manage the invasive insect.

7. Conclusion

Globalisation has enriched us beyond belief, but it too comes with a consequence. Increased connectivity due to globalisation has resulted in the introduction of numerous invasive insect pests to India which not only have devastated several agricultural crops but also have caused huge monetary loss. This situation further got worse with the advent of global warming and climate change which favoured the establishment and spread of the invasive species. Hence, management of the invasive insects is a challenge. Minimising the loss caused by invasive insects requires an "international management approach" with strict legislation laws and better cooperation among countries with respect to exchange of information regarding the invasive species and their natural enemies. Moreover, there should be a hierarchical setup recruiting expertise personnel having sound knowledge on insect identification, preliminary risk assessment and monitoring of insects and their eradication. Besides, several public awareness campaigns can be conducted to educate the common people which will definitely reduce the chances of an accidental invasion.

Author details

Swoyam Singh*, Jai Hind Sharma, Avinash Udikeri and Hashib Ansari
GB Pant University of Agriculture and Technology, Uttarakhand, India

*Address all correspondence to: swoyamsingh@gmail.com

References

[1] Stanley AW, Sanjay S, et al. India [Internet]. 2020. Available from: https://www.britannica.com/place/India

[2] Ahmad L, Kanth RH, Parvaze S, Mahdi SS. Agro-climatic and agro-ecological zones of India. In: Experimental Agrometeorology: A Practical Manual. Cham: Springer; 2017. pp. 99-118

[3] Wikipedia [Internet]. Available from: https://en.wikipedia.org/wiki/Geography_of_India

[4] Gupta N, Verma SC, Sharma PL, Thakur M, Sharma P, Devi D. Status of invasive insect pests of India and their natural enemies. Journal of Entomology and Zoology Studies. 2019;7(1):482-489

[5] Reshi ZA, Khuroo AA. Alien plant invasions in India: Current status and management challenges. In: Proceedings of the National Academy of Sciences, India - Section B: Biological Sciences. 2012;(82):305-312

[6] Sandilyan S. Invasive Species of India Report. National Biodiversity Authority, Ministry of Environment Forests and Climate Change Government of India [Internet]. Available from: http://nbaindia.org/uploaded/pdf/Iaslist.pdf

[7] Singh S, Mukherjee SK. Studies on flowering thrips of cashew and its correlation on nut shedding. Journal of Entomology and Zoology Studies. 2018;**6**(4):873-875

[8] Piel F, Gilbert M, De Cannière C, Grégoire JC. Coniferous round wood imports from Russia and Baltic countries to Belgium. A pathway analysis for assessing risks of exotic pest insect introductions. Diversity and Distributions. 2008;**14**(2):318-328

[9] Kaur M, Aggarwal NK, Kumar V, Dhiman R. Effects and management of *Parthenium hysterophorus*: A weed of global significance. International Scholarly Research Notices. 2014;**2014**:1-12

[10] Gupta R. Management and comparative efficacy of various treatments against *Eriosoma lanigerum* HAUSMANN on apple trees (*Malus domestica* BORKH.) in Jammu province of Jammu & Kashmir state. International Journal of Advanced Biological Research. 2015;**5**(4):319-321

[11] India tourism statistics at a glance. Government of India [Internet]. 2018. Available from: http://tourism.gov.in/sites/default/files/Other/India%20Tourism%20Statistics%20at%20a%20Glance%202019.pdf

[12] Kiritani K, Yamamura K. Exotic insects and their pathways for invasion. In: Invasive Species: Vectors and Management Strategies. Washington: Island Press; 2003. pp. 44-67

[13] Work TT, McCullough DG, Cavey JF, Komsa R. Arrival rate of nonindigenous insect species into the United States through foreign trade. Biological Invasions. 2005;7(2):323

[14] Dobbs TT, Brodel CF. Cargo aircraft as a pathway for the entry of nonindigenous pests into south Florida. Florida Entomologist. 2004;**87**(1):65-78

[15] Hulme PE. Trade, transport and trouble: managing invasive species pathways in an era of globalization. Journal of Applied Ecology. 2009;**46**(1):10-18

[16] Yan X, Zhenyu L, Gregg WP, Dianmo L. Invasive species in China—An overview. Biodiversity and Conservation. 2001;**10**(8):1317-1341

[17] Anderson LG, Rocliffe S, Haddaway NR, Dunn AM. The role of

tourism and recreation in the spread of non-native species: A systematic review and meta-analysis. PLoS One. 2015;**10**(10):1-15

[18] Kolar CS, Lodge DM. Ecological predictions and risk assessment for alien fishes in North America. Science. 2002;**298**(5596):1233-1236

[19] Osyczka P, Mleczko P, Karasiński D, Chlebicki A. Timber transported to Antarctica: A potential and undesirable carrier for alien fungi and insects. Biological Invasions. 2012;**14**(1):15-20

[20] Johnson SJ. Migration and the life history strategy of the fall armyworm, *Spodoptera frugiperda* in the Western Hemisphere. International Journal of Tropical Insect Science. 1987;**8**(4-6):543-549

[21] Bajracharya AS, Bhat B, Sharma P, Shashank PR, Meshram NM, Hashmi TR. First record of fall army worm *Spodoptera frugiperda* (JE Smith) from Nepal. Indian Journal of Entomology. 2019;**81**(4):635-639

[22] Nail KR, Drizd L, Voorhies KJ. Butterflies across the globe: A synthesis of the current status and characteristics of monarch (*Danaus plexippus*) populations worldwide. Frontiers in Ecology and Evolution. 2019;7:362

[23] Diez JM, D'Antonio CM, Dukes JS, Grosholz ED, Olden JD, Sorte CJ, et al. Will extreme climatic events facilitate biological invasions? Frontiers in Ecology and the Environment. 2012;**10**(5):249-257

[24] Kelley AL. The role thermal physiology plays in species invasion. Conservation Physiology. 2014;**2**(1):1-14

[25] Pachauri RK, Allen MR, Barros VR, Broome J, Cramer W, Christ R, et al. Climate Change 2014: Synthesis Report. Contribution of Working Groups I, II and III to the Fifth Assessment Report of the Intergovernmental Panel on Climate Change. Ipcc; 2014

[26] Bonte D, Baert L, Lens L, Maelfait J-P. Effects of aerial dispersal, habitat specialisation, and landscape structure on spider distribution across fragmented grey dunes. Ecography. 2004;**27**(3):343-349

[27] Renault D, Laparie M, McCauley SJ, Bonte D. Environmental adaptations, ecological filtering and dispersal central to insect invasions. Annual Review of Entomology. 2018;**63**:345-368

[28] Mishra SC. The American blight or wooly apple aphid *E. Lanigerum* (Hausmann). Agricultural Research Journal. 1920;**15**:627

[29] Singh SP. Some Success Stories in Classical Biological Control of Agricultural Pests in India. Asia-Pacific Association of Agricultural Research Institutions (APAARI). Bankok, Thailand: FAO; 2004

[30] Muniappan R, Viraktamath CA. Status of biological control of the weed, Lantana camara in India. International Journal of Pest Management. 1986;**32**(1):40-42

[31] Jalali SK, Singh SP. Release and recovery of an exotic coccinellid predator, *Curinus coeruleus* (Muls.) on subabul psyllid, *Heteropsylla cubana* Crawf. in India. Journal of Insect Science. 1989;**2**(2):158-159

[32] Palaniswami MS, Pillai KS, Nair RR, Mohandas C. A new cassava pest in India. Cassava News. 1995;**19**(1):6-7

[33] Jhala RC, Bharpoda TM, Patel MG. *Phenacoccus solenopsis* Tinsley (Hemiptera: Pseudococcidae), the mealy bug species recorded first time on cotton and its alternate host plants in Gujarat, India. Uttar Pradesh Journal of Zoology. 2008;**28**(3):403-406

[34] Nagrare VS, Kranthi S, Biradar VK, Zade NN, Sangode V, Kakde G, et al. Widespread infestation of the exotic mealybug species, *Phenacoccus solenopsis* (Tinsley) (Hemiptera: Pseudococcidae), on cotton in India. Bulletin of Entomological Research. 2009;**99**(5):537-541

[35] Abbas G, Arif MJ, Ashfaq M, Aslam M, Saeed S. Host plants, distribution and overwintering of cotton mealybug(*Phenacoccus solenopsis*; Hemiptera: Pseudococcidae). International Journal of Agriculture and Biology. 2010;**12**(3):421-425

[36] Srinivasan T, Saravanan PA, Josephrajkumar A, Rajamanickam K, Sridharan S, David PM, et al. Invasion of the Rugose spiralling whitefly, *Aleurodicus rugioperculatus* Martin (Hemiptera: Aleyrodidae) in Pollachi tract of Tamil Nadu, India. Madras Agriculture Journal. 2016;**103**(10-12):349-353

[37] NBAIR [Internet]. Available from: https://www.nbair.res.in/sites/default/files/2019-07/pest%20alert%2023%20july%202019.pdf

[38] NBAIR [Internet]. Available from: https://nbair.res.in/sites/default/files/2019-10/Pest%20Alert%20Aleurothrixus%20floccosus.pdf

[39] Fletcher TB. Some south Indian Insects and Other Animals of Importance Considered Especially from an Economic Point of View. Printed by the Superintendent, Government Press, Madras; 1914

[40] Sridhar V, Chakravarthy AK, Asokan R, Vinesh LS, Rebijith KB, Vennila S. New record of the invasive South American tomato leaf miner, *Tuta absoluta* (Meyrick) (Lepidoptera: Gelechiidae) in India. Pest Management in Horticultural Ecosystem. 2014;**20**(2):148-154

[41] Shylesha AN, Jalali SK, Gupta AN, Varshney RI, Venkatesan T, Shetty PR, et al. Studies on new invasive pest *Spodoptera frugiperda* (JE Smith) (Lepidoptera: Noctuidae) and its natural enemies. Journal of Biological Control. 2018;**32**(3):145-151

[42] Faizal MH, Prathapan KD, Anith KN, Mary CA, Lekha M, Rini CR. Erythrina gall wasp *Quadrastichus erythrinae*, yet another invasive pest new to India. Current Science. 2006;**90**(8):1061-1062

[43] Varaprasad KS, Balasubramanian S, Diwakar BJ, Rao CR. First report of an entomogenous nematode, *Panagrolaimus sp*. from coffee-berry borer, *Hypothenemus hampei* (Ferrari) from Karnataka, India. Plant Protection Bulletin. 1994;**46**(2/3):34

[44] Shashank PR, Suroshe SS, Singh PK, Chandrashekar K, Nebapure SM, Meshram NM. Report of invasive tomato leaf miner, *Tuta absoluta* (Lepidoptera: Gelechiidae) from northern India. Indian Journal of Agricultural Sciences. 2016;**86**(12):1635-1636

[45] Plant Quarantine India [Internet]. Available from: http://plantquarantineindia.nic.in/PQISPub/docfiles/dip_act.htm

[46] MHFW. The Gazette of India. Notification.1037, 1-3. 2017. Available from: http://www.cdsco.nic.in/writereaddata/GSR327(E) Dated 03_04_2017.pdf

[47] DPPQS [Internet]. Available from: http://ppqs.gov.in/divisions/plant-quarantine/strengthening-modernisation-plant-quarantine-facilities-india

[48] Sreedevi K, Meshram N, Shashank PR. Insect Taxonomy- Basics to barcoding. In: Chakravarthy AK, editor. New Horizons in Insect Science: Towards Sustainable Pest Management.

India: Springer; 2015:3-12. DOI: 10.1007/978-81-322-2089-3

[49] Station AB-R, Cargo A. Plant Quarantine Including Internal Quarantine Strategies in View of Onslaught of Diseases and Insect Pests. New Delhi, India: NAAS; 2003. p. 1067

[50] David P, Thebault E, Anneville O, Duyck PF, Chapuis E, Loeuille N. Impacts of invasive species on food webs: A review of empirical data. Advances in Ecological Research. 2017;**56**:1-60

[51] Kenis M, Hurley BP, Colombari F, Lawson S, Sun J, Wilcken C, et al. Guide to the classical biological control of insect pests in planted and natural forests. FAO Forestry Papers. 2019;**182**:1-3

[52] Barratt BI, Moran VC, Bigler F, Van Lenteren JC. The status of biological control and recommendations for improving uptake for the future. BioControl. 2018;**63**(1):155-167

[53] Babendreier D. Pros and cons of biological control. In: Biological Invasions. Berlin, Heidelberg: Springer; 2008. pp. 403-418

[54] Siddhapara MR, Patel MB, Patel HV. Biology of *Zygogramma bicolorata* Pallister (Coleoptera:Chrysomelidae) and their feeding potential on Parthenium and sunflower. The Madras Agricultural Journal. 2012;**99**(10/12):841-844

Chapter 4

The Spotted Wing Drosophila in the South of the World: Chilean Case and Its First Productive Impacts

Karina Buzzetti Morales

Abstract

The spotted wing drosophila *Drosophila suzukii* Matsumura (Diptera: Drosophilidae) is endemic in Asia. This species was first reported in Chile in early 2017 in the region of La Araucanía, Los Ríos, and Los Lagos, but it has dispersed to other regions of the south-central area of the country, currently being in the category of plague present with restricted distribution. *D. suzukii* is a polyphagous pest, and it infests a wide range of fruit crops, including grapes, cherries, as well as an increasing number of wild fruits. Chile is the main cherry-producing country in the south of the world, providing mainly to the Asian market, so, although the pest is not a quarantine problem for this destination of the fruit, its potential damage is high due to the decrease in yield, fruit deterioration, and increased pre- and postharvest rot. The local productive reality also provides various hosts to the plague, allowing its expression and adaptation, for example, in blueberry orchards. This report summarizes the first signs of economic damage and explores the first results of control delivered by various insecticides: alpha-cypermethrin, acetamiprid, chlorantraniliprole, cyantraniliprole, emamectin benzoate, lambda- and gamma-cyhalothrin, spinetoram, and spinosad. Also, repellent effect of other compounds was studied.

Keywords: fruit fly, insecticides, Drosophilidae, invasive pest, economic losses

1. Introduction

Drosophila suzukii Matsumura (Diptera: Drosophilidae) is classified by the European and Mediterranean Plant Protection Organization (EPPO) as an emerging and invasive pest [1]. This insect is native to east and southeast Asia, although its real geographical origin is still discussed [2]. It has been introduced in several Hawaiian Islands [2], North America [3], Europe [4], and recently in Iran [5], Argentina [6], and later Chile [7], reflecting its high adaptive capacity. Its relevance is due to high economic importance because of the severity of the damage it causes in various productive crops [8–15]. It is a very polyphagous pest, with more than 50 hosts reported in literature [16].

In contrast with *Drosophila melanogaster*, the spotted wing fruit fly develops in mature and undamaged fruits, injecting its eggs into fruits when they preferably begin to turn color [17]. The larvae develop inside the fruits, finally pupating on or inside them [15, 17]. If there is not enough healthy fruit, it can also attack fallen or damaged fruit [18].

Since only 2 years ago its presence was reported in Chile, until now there was no corroboration of its productive impact in the country. At the same time, for the design of the management proposals, there was only external literature; however, the objective of this work was to monitor the economic damage caused by the plague in cherry and blueberry orchards and also evaluate the control achieved with insecticides of different chemical groups: alpha-cypermethrin, acetamiprid, chlorantraniliprole, cyantraniliprole, emamectin benzoate, lambda- and gamma-cyhalothrin, spinetoram, and spinosad.

The possible repellent effect of garlic and chili-based products, as well as an extract of *Chenopodium ambrosioides*, was also investigated.

2. Methodology

2.1 Fruit damage monitoring

Five commercial orchards were selected from localities Cachapoal and Culenar, Ñuble Region, in areas with positive *D. suzukii* detections since 2017. From the newly formed fruit state, in September 2019, cherry orchards and blueberry orchards were monitored three times per week until January 2020. Production areas of more than 10 hectares were addressed and with at least 10 years of operation. The orchards correspond to conventional production, with drip technified irrigation and with similar agricultural management conditions with each other.

Three varieties of cherries (Lapins, Santina, and Regina) and two of blueberries (Brigitta and Elliot), as also material from live fences of *Aristotelia chilensis*, *Rubus ulmifolius*, and *Prunus cerasifera*, were included in this monitoring. From each substrate and on each occasion, 100 random fruits were collected, and some insect individuals were selected for analysis with traditional taxonomy and/or PCR analysis for the larva and pupa states extracted from the samples. The monitoring record was carried in parallel to the climate record of temperature and relative humidity.

At harvest, losses associated with the pest were estimated in orchard and prepackaged fruit inspection.

2.2 Efficacy test

Once the start of the attack of *D. suzukii* on fruits was determined, the cherry orchard with the highest severity of attack was selected to perform efficacy studies. In this step, a cherry orchard of cv. Regina (1200 plants/ hectare) was used. Insecticides from five chemical groups were compared [19]: group 3A (pyrethroids: alpha-cypermethrin and lambda- and gamma-cyhalothrin), group 4A (neonicotinoids: acetamiprid), group 5 (spinosyns: spinetoram and spinosad), group 6 (avermectins: emamectin benzoate), and group 28 (diamides: chlorantraniliprole and cyantraniliprole).

The treatments evaluated were Mageos® (15% w/w alpha-cypermethrin dispersible granulated formulation; 15 g of commercial product/100 L) (Basf Chile S.A); Bull® (6.12% w/v gamma-cyhalothrin encapsulated suspension; 10 mL of commercial product/100 L) (FMC Chemical Chile Ltd); Karate® Zeon (5% w/v lambda-cyhalothrin encapsulated suspension; 20 mL of commercial product/100 L) (Syngenta S.A); Mospilan® (20% w/w acetamiprid soluble powder; 50 g/100 L) (Basf Chile S.A); Success® 48 (45% w/v spinosad concentrated suspension;

14 g/100 L); Delegate® (25% w/w spinetoram concentrated suspension; 16 g/100 L); Proclaim® Forte (5% w/w emamectin benzoate dispersible granulate: 30 g/100 L) (Syngenta S.A); Coragen® (20% w/v chlorantraniliprole concentrated suspension; 20 mL/100 L) (FMC Chemical Chile Ltd); and Exirel® (10% w/v cyantraniliprole suspoemulsion; 75 mL/100 L) (FMC Chemical Chile Ltd). A control without insecticide was included.

A completely randomized design was used, with four repetitions of each treatment. The experimental unit corresponded to groups of 10 plants of the same row treated with 4 isolation plants between them.

All applications were performed once, on November 15 (when fruit begins to take color), with a conventional hydraulic sprayer (Line Patasa 2000) with 1800 L of water per hectare. A control treatment without insecticide applications was considered.

Two parameters of relevance were evaluated: the incidence of damage (established as mean of fruits damaged by the pest) and the severity of the damage (wound or presence of live larvae) on samples of 100 fruits collected randomly within each experimental unit. Fruits were reviewed at 3, 5, 7, 10, 12, and 14 days after application (DAA). The cutoff criterion to stop evaluations was unified when the incidence of damage was equal to or greater than 20%.

The data obtained from field and laboratory measurements was subjected to analysis of variance (ANOVA) by taking appropriate transformations. Mean comparisons in significant ANOVAs were performed with a Tukey's test. Statistical analyses were conducted using the software Minitab®16.1.0 (Minitab Inc.).

2.3 Repellency test

In a cherry orchard of var. Regina located in the town of Culenar, the possible effect of repellency (expressed as a lower incidence of eggs laid of *D. suzukii* on cherries) achieved by the use of two commercial products based on natural compounds, Amarex® (Captiva® Prime) which is formulated by 7.60% v/v capsicum oleoresin extract, 23.40% v/v garlic oil, and 55% v/v canola oil (Gowan Chile SpA) and Requiem® Prime which is formulated by 9.1% w/v α-terpinene, 3.4% w/v p-cymene, and 2.73% w/p d-limonene (Bayer S.A), was studied. Both were applied three times every 5 days as fruit starts to change color.

Treatment with Amarex® was applied at 200 mL/100 L, and treatment with Requiem® Prime was applied at 240 mL/100 L, using a conventional hydraulic sprayer (Line Patasa 2000) with 1800 L of water per hectare. One check control using only water treatment was included.

All treatment had four repetitions of 10 plants, randomly drawn inside the orchard. Evaluations were performed at 3 days between applications and at 3 and 7 days after the last application.

On each occasion the frequency of egg of *D. suzukii* laid in fruits was evaluated, in samples of 100 fruits collected at random.

After each evaluation involved in the present work, the destruction of the contaminated material was carried out via freezing at -30C for 5 days.

3. Results

All monitored hosts showed damage and development of immature states of the pest from the beginning of fruit maturity. The above adds *A. chilensis*, a Chilean native plant, and *Prunus cerasifera* (**Figure 1**) to the list of secondary hosts of the pest.

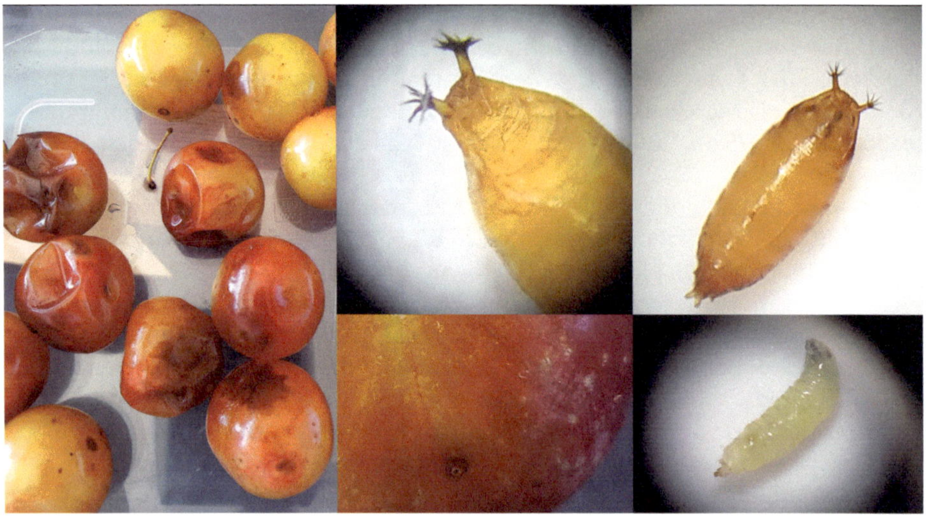

Figure 1.
Inmature stages of D. suzukii *collected from fruits of* Prunus cerasifera.

The period of greatest incidence of the pest was at the beginning of fruit color breakage and increases as it approaches maturity. Development persists in remnants of the orchard after harvest (**Figures 2** and **3**).

There were no indications of preference of attack of one fruit variety over another nor signs of preference among fruit species when comparing the incidence between cherry trees and blueberries in similar conditions. The foregoing could vary according to different pesticides or nutritional programs or driving management used among these crops, but at least the records collected suggest that similar precautions should be handled in these fruit trees.

There was no rainfall in the period of pest incidence, and daily temperatures ranged from 6 to 36°C. Relative humidity varied from 40 to 65%, that is, adjusted to favorable climatic requirements for the development of the pest [20]. The area of cherries in Chile currently exceeds 27,000 hectares, concentrated in areas that have favorable climatic conditions for this pest, a situation that is even more marked in the case of blueberry production.

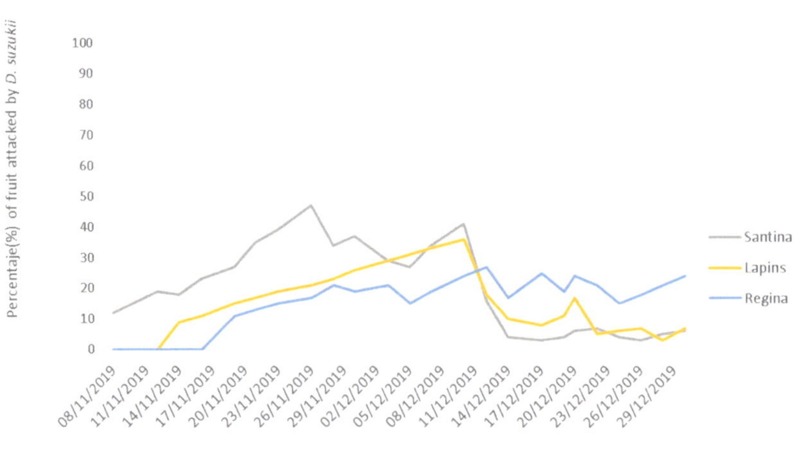

Figure 2.
Percentage (%) of cherry fruits attacked by D. suzukii *separated by cultivar.*

Figure 3.
Percentage (%) of blueberry fruits attacked by D. suzukii *separated by cultivar.*

Since Chile is a country focused on the export of fruits, the damages observed until now can be categorized in two categories: complete loss of the fruit (**Figure 4**) or loss of quality due to deterioration or damage to the skin of the fruit (**Figure 5**).

In the monitored cases, for cherries, the first category reached between 10% and15% of the production (approximately 10% Santina, 12% Lapins, and 15% Regina), that is, the equivalent of 1, 2.16, and 2.7 ton/ha. Economically speaking, it is a loss equivalent to 5,000–17,550 USD/hectare. The second item is still difficult to quantify. The second item was estimated at 1–2%.

In the case of blueberries, the percentage of compromised fruit was close to 8%, which could increase as the plague increases its local population. This corresponds to the equivalent of 1–1.5 ton/ha or a loss equivalent to 4,000 USD/hectare.

In fruits of live fences (*A. chilensis*, *R. ulmifolius*, and *P. cerasifera*), at the end of the season, the recorded frequency was greater than 60%, possibly since they do not receive pesticide treatments and not necessarily because there is a greater attraction or preference to those substrates. Unfortunately, at least two of these three species are widely distributed in the country, such as ornamental plants or live fences, which contribute to the spread of the pest.

Figure 4.
Losses caused in cherry trees by massive attack of D. suzukii.

Figure 5.
Cherries showing damage by egg laying attempts.

Secondary damage due to attempts to lay eggs occurred in some sectors that received certain insecticidal treatments and allowed the fruit to be redirected to the domestic market, although at a significative lower return value than expected in an export fruit. Concordant with what happened in other countries, greater complexity is expected from the point of view of the increase of residues present in fruit due to the management of this pest [21].

Control delivered by the microencapsulated pyrethroid treatments, both spinosyns, and cyantraniprole was highly efficient not only in reducing the incidence of infestation (**Figure 6**) but also in decreasing the severity of the attack (**Figure** 7).

Spinosyns and some pyrethroids such as l-cyhalothrin are indicated with high efficacy in the control of damage of this pest, at short intervals of application to avoid egg postures [22].

In the case of acetamiprid, emamectin benzoate, and chlorantraniliprole, these showed weaker insecticidal action than the previous group, considering that the control action was only expressed once the fruit was affected by the pest. Even so, because they show a significant decrease in the incidence of damage for an approximate period of 5 days, future work may focus on complementing these mechanisms of action with others (such as repellents) that allow reducing the severity of the damage. These results are consistent with what was raised by specialists in the United States [23]. Nevertheless, despite being promising in management, the high rate of dispersal, reproduction, and adaptation of the pest is consistent with the high risk of economic damage reported globally [24].

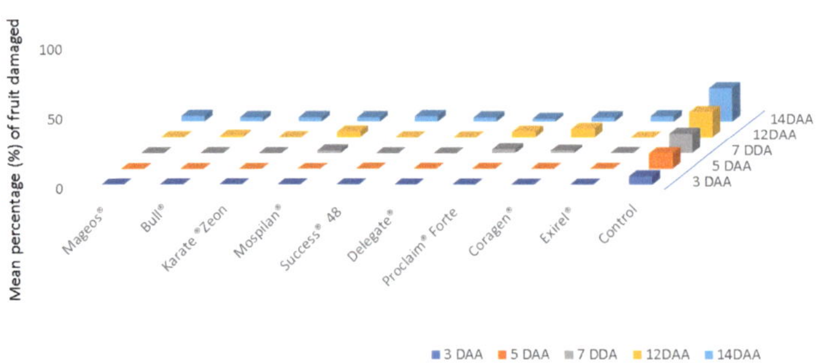

Figure 6.
Mean percentage (%) of fruits damaged by D. suzukii *days after each insecticide treatment.*

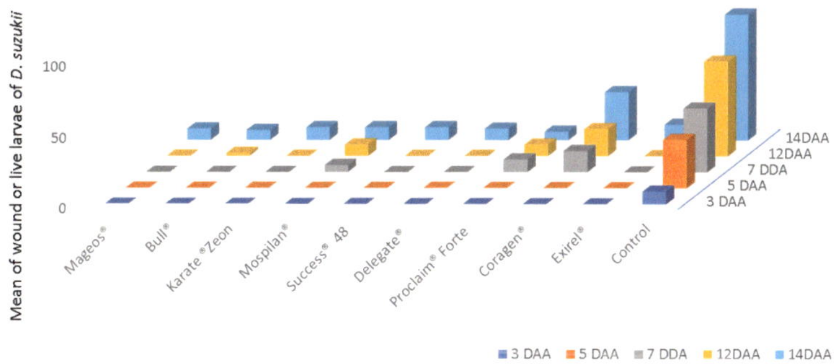

Figure 7.
Mean of wound/or live larvae of D. suzukii *in fruits days after each insecticide treatment.*

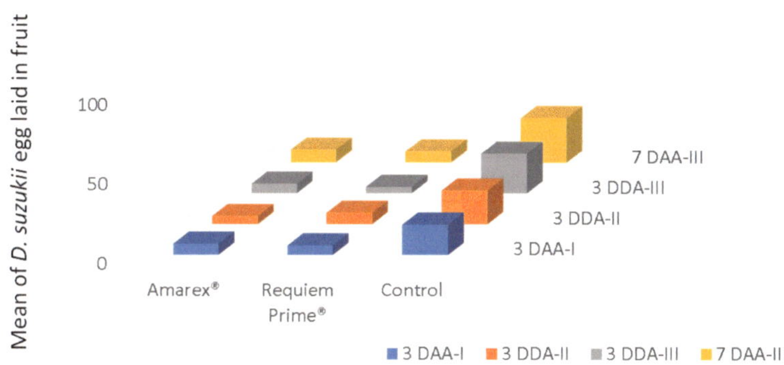

Figure 8.
Severity of attack (frequency of D. suzukii *egg position).*

On the other hand, although the history of efficient biological control has been weak [23], several authors agree on the need to advance in the integral knowledge of the management of this insect, including predators, parasitoids, repellents, insecticides, and cultural management [25–30].

Regarding the effect granted using products based on natural compounds, both Requiem® Prime and Amarex® demonstrated at least the decrease in the severity of attack of the pest under conditions of use in the orchard (**Figure 8**). However, it should be considered that the only use of these treatments may not be enough to reduce the economic impact when faced with increased *D. suzukii* pressure. More studies should be realized about their possible contribution inserted in a program or in mixture with conventional insecticides in order to contribute to complement mode of action.

The mode of action as insecticide reported for Requiem® Prime are explained because the active ingredient cause degradation of soft insect cuticles results in the disruption of insect mobility and respiration [31]. In the case of Amarex®, its action would also be explained by destruction of membranes added to the repellent effect [32].

4. Conclusions

The results of the monitoring of damage in fruits and its evolution during the season reflect that the potential risk of the plague in a country with the productive reality of Chile is high and that locally *D. suzukii* has found new shelters that allow it to spread and maintain its development. Although, in the search for its control, adequate tools were determined to reduce its damage, a greater understanding of the mechanism of action of each product is required in order to position it properly, in order to reduce the potential damage of the insect.

Given the high-quality requirement presented by the export fruit, there is a high risk of selection of less sensitive individuals of the pest, and those cases of violations of the maximum limits of pesticide residues occur due to an overheating of applications. For this reason, future work should consider other integrated management edges. In this line, it is necessary to investigate locally the use of possible mixtures of repellents with insecticides, the use of biological controllers in parallel to a management program, and the use of mass capture traps.

For now, two formulated products (Requiem® Prime and Amarex®) have promising results to achieve a lower severity of damage without adding residues to the fruit, which can be complementary to the use of insecticides aimed primarily at the management of adults of this pest.

Of the insecticides compared, the control delivered by cyantraniliprole and both spinosyns stands out. All microencapsulated pyrethroids showed stable control for at least 7 days. Acetamiprid, emamectin benzoate, and chlorantraniliprole were not efficient in reducing the attempt to lay eggs but were enough to decrease the incidence of damage. Therefore, in the short term, in Chile there are adequate tools for the management of this insect, but adequate use should be provided in order to preserve food safety.

Caution should also be taken with the selection of less sensitive individuals of the pest, because, due to the characteristics of the species, the expression of resistance to insecticides can enhance their economic damage.

Acknowledgements

The author personally thanks Juan Carlos Ríos, Ricardo Larral, Alejandro Salas M., and Fernando Jofre. Acknowledgments are also expressed to SAG officials Rodrigo Astete, Marco Muñoz, Ruth Castro, Claudio Moore T., Nayadeth Álvarez, Oscar Rojas, and Marta Solís.

The study was funded exclusively by AgriDevelopment Ltd.

Conflict of interest

The author declares no conflict of interest in the delivery information.

Author details

Karina Buzzetti Morales
Consultant AgriDevelopment Ltd, Santiago, Chile

*Address all correspondence to: karinabuzzetti@agridev.cl

References

[1] European and Mediterranean Plant Protection Organization (EPPO). Global Database. EPPO A2 List of pests recommended for regulation as quarantine pests (version 19/09). Available from: https://www.eppo.int/ACTIVITIES/plant_quarantine/A2_list [Accessed: 05 January 2020]

[2] Hauser M. A historic account of the invasion of Drosophila suzukii (Matsumura) (Diptera: Drosophilidae) in the continental United States, with remarks on their identification. Pest Management Science. 2011;**67**(11):1352-1357

[3] Burrack HJ, Smith JP, Pfeiffer DG, Koeher G, Laforest J. Using volunteer-based networks to track Drosophila suzukii (Diptera: Drosophilidae) an invasive pest of fruit crops. Journal of Integrated Pest Management. 2012;**3**(4):B1-B5

[4] Cini A, Ioriatti C, Anfora G. A review of the invasion of Drosophila suzukii in Europe and a draft research agenda for integrated pest management. Bulletin of Insectology. 2012;**65**(1):149-160

[5] Mehrdad P-A, Gilasian E, Kayhanian AA. Spotted wing drosophila, Drosophila suzukii (Matsumura) (Dip.: Drosophilidae), an invasive fruit pest new to the Middle East and Iran. Drosophila Information Service. 2015;**98**:59-60

[6] ChiaHua L, Mottern JL, Walsh GC, Buffington ML. New record for the invasive spotted wing drosophila, Drosophila suzukii (Matsumura, 1931) (Diptera: Drosophilidae) in Anillaco, western Argentina. Proceedings of the Entomological Society of Washington. 2017;**119**(1):146-150

[7] Servicio Agrícola y Ganadero (SAG) Chile. Plagas relevantes presentes. Drosophila suzukii. 2019. Available from: https://www.sag.gob.cl/ambitos-de-accion/drosophila-suzukii [Accessed: 05 January 2020]

[8] Grassi A, Palmie L, Giongo L. New pests of the small fruits in Trentino. (Nuovo fitofago per i piccoli frutti in Trentino.). Terra Trentina. 2009;**55**(10):19-23

[9] Bolda M, Goodhue RE, Zalom FG. Spotted wing drosophila: Potential economic impact of a newly established pest. Agricultural and Resource Economics. 2010;**13**(3):5-8

[10] Goodhue RE, Bolda M, Farnsworth D, Williams JC, Zalom FG. Spotted wing drosophila infestation of California strawberries and raspberries: Economic analysis of potential revenue losses and control costs. Pest Management Science. 2011;**67**(11):1396-1402

[11] De Ros G, Anfora G, Grassi A, Ioriatti C. The potential economic impact of Drosophila suzukii on small fruits production in Trentino (Italy). IOBC-WPRS Bulletin. 2013;**91**:317-321

[12] Mazzi D, Bravin E, Meraner M, Finger R, Kuske S. Economic impact of the introduction and establishment of Drosophila suzukii on sweet cherry production in Switzerland. Insects. 2017;**8**(1):18

[13] De Ros G, Conci S, Pantezzi T, Savini G. The economic impact of invasive pest Drosophila suzukii on berry production in the province of Trento, Italy. Journal of Berry Research. 2015;**5**(2):89-96

[14] Wiman NG, Dalton DT, Anfora G, Biondi A, Chiu JC, Daane KM, et al. Drosophila suzukii population response to environment and management strategies. Journal of Pest Science. 2016;**89**(3):653-665

[15] Kanzawa T. Studies on Drosophila suzukii Mats. Kofu, Yamanashi agricultural Experiment Station. Abstract in the Review of Applied Entomology. 1939;**29**:622

[16] Centre for Agricultural Bioscience International (CABI). D. suzukii, host/species affected. 2020. Available from: https://www.cabi.org/isc/datasheet/109283#tohostsOrSpeciesAffected [Accessed: 05 January 2020]

[17] Woltz JM, Lee JC. Pupation behavior and larval and pupal biocontrol of Drosophila suzukii in the field. Biological Control. 2017;**110**:62-69

[18] Burrack HJ, Fernandez GE, Spivey T, Kraus DA. Variation in selection and utilization of host crops in the field and laboratory by Drosophila suzukii Matsumura (Diptera: Drosophilidae), an invasive frugivore. Pest Management Science. 2013;**69**(10):1173-1180

[19] Insecticide Action Committee (IRAC). 2019. IRAC Mode of Action Classification Scheme, Issued June 2019. p. 30

[20] Tochen S, Dalton DT, Wiman NG, Hamm C, Shearer PW, Walton VM. Temperature-related development and population parameters for Drosophila suzukii (Diptera: Drosophilidae) on cherry and blueberry. Environmental Entomology. 2014;**43**:501-510

[21] Haviland DR, Beers EH. Chemical control programs for Drosophila suzukii that comply with international limitations on pesticide residues for exported sweet cherries. Journal of Integrated Pest Management. 2012;**3**(2):F1-F6

[22] Torné M, Abad R, Jacquet V. Spinosines®, una herramienta imprescindible para el control de *Drosophila suzukii* Matsumara. Phytoma. 2014;**265**:26-28

[23] Asplen MK, Anfora G, Biondi A, Choi DS, Chu D, Daane KM, et al. Invasion biology of spotted wing drosophila (Drosophila suzukii): A global perspective and future priorities. Journal of Pest Science. 2015;**88**(3):469-494

[24] Bolda MP, Goodhue RE, Zalon FG. Spotted wing drosophila: Potential economic impact of a newly established pest. The Giannini Foundation of Agricultural Economics. 2010;**13**:5-8

[25] Walsh DB, Bolda MP, Goodhue RE, Dreves AJ, Lee J, Bruck DJ, et al. Drosophila suzukii (Diptera: Drosophilidae): Invasive pest of ripening soft fruit expanding its geographic range and damage potential. Journal of Integrated Pest Management. 2011;**2**:G1-G7

[26] Schlesener D, Wollmann J, Pazini J, Grützmacher A, Mello Garcia F. Effects of insecticides on adults and eggs of Drosophila suzukii (Diptera, Drosophilidae). Revista Colombiana de Entomologia. 2017;**43**(2):208-214

[27] Rossi M, Grassi A, Dalton DT, Miller B, Ouantar M, Loni A, et al. First field records of Pachycrepoideus vindemiae as a parasitoid of Drosophila suzukii in European and Oregon small fruit production areas. Entomologia. 2013;**1**:e3

[28] Rossi M, Buffington M, Daane KM, Dalton DT, Grassi A, Kacar G, et al. Host stage preference, efficacy and fecundity of parasitoids attacking Drosophila suzukii in newly invaded areas. Biological Control. 2015;**84**:28-35

[29] Caprile J, Flint ML, Bolda MP, Grant JA, Van Steenwyk R, Haviland D. 2013. Spotted wing Drosophila (Drosophila suzukii). University of California-IPM online. Available from: http://www.ipm.ucdavis.edu/EXOTIC/drosophila.html. [Accessed: 07 January 2020]

[30] Cormier D, Veilleux J, Firlej A. Exclusion net to control spotted wing drosophila in blueberry fields. IOBC-WPRS Bulletin. 2015;**109**:181-184

[31] Certis USA 2019. Requiem® Prime. Available from: https://www.certisusa.com/pest_management_products/bioinsecticides/requiemprime [Accessed: 05 January 2020]

[32] Gowan Company. 2020. Captiva® Prime. Available from: https://www.gowanco.com/products/captiva-prime [Accessed: 11 January 2020]

Chapter 5

Tuta absoluta (Meyrick) (Lepidoptera: Gelechiidae): An Invasive Insect Pest Threatening the World Tomato Production

Hamadttu Abdel Farag El-Shafie

Abstract

The South American tomato pinworm or tomato leaf miner (TLM), *Tuta absoluta* (Meyrick) (Lepidoptera: Gelechiidae), is a serious invasive and destructive insect pest of tomato (*Solanum lycopersicum* L.) worldwide. The moth can cause 100% damage in tomato crop in both greenhouses and open fields if control measures are not carried out. Due to the high reproduction potential, dispersal ability, and tolerance to environmental conditions, the TLM invaded most tomato-producing countries in Europe, Africa, and Asia. The tomato leaf miner originated in South America and was first introduced in Spain in 2006 and from where it spread to other part of the world. This chapter consolidates the rich literature on the pest with emphasis on invasion history, economic significance, and possible management options adopted worldwide.

Keywords: tomato leafminer, *Solanum lycopersicum*, invasive potential, management, Gelechiidae, quarantine pest

1. Introduction

Biological invasion has occurred for millennia, but increased globalization in recent decades has accelerated it [1]. Invasive insect species reduce crop yield, increase cost of production especially pest control costs, increase reliance on pesticides, and disrupt preexisting integrated pest management (IPM) programs. Invasive insect species cause considerable damage to agriculture, horticulture, and forest industries worldwide [2, 3] with an estimated annual economic loss of about 70 billion US$ [4]. Transportation and international trade are increasing rapidly, thus facilitating the spread and dispersal of invasive species [5]. The tomato (*Solanum lycopersicum* L.) is an important horticultural vegetable crop that is only second to potato. The total world production of tomato is about 180 million tons grown in areas of approximately 4 M ha. The top 10 tomato-producing countries in the world are China, India, USA, Turkey, Egypt, Italy, Iran, Spain, Brazil, and Mexico. China, India, and Turkey, account for almost half of the land area covered worldwide with tomato crops, that is, 31, 11, and 7%, respectively [6]. Tomato is the sixth most valuable cultivated crop in the world worth US$ 87.9 billion in 2016. The tomato leaf miner *Tuta absoluta* (Lepidoptera: Gelechiidae) is threatening about 87% of this production worldwide [3, 6, 7]. *T. absoluta* has several common English names in the literature. These are the South American tomato pinworm, the South American tomato leaf miner, the South American tomato moth, the tomato pinworm, the tomato borer, and the tomato leaf miner. For consistency, the tomato leaf miner (TLM) will be used throughout this chapter. The TLM has been considered as a key pest of tomato, in recent years, causing a reduction in tomato yield that can reach 100% if no management action is taken [1]. Increasing of global trade of tomato in the absence of strict quarantine measures and proper surveillance in many tomato-producing countries are the reason behind the vast spread of this pest.

Due to the significance of TLM, the Journal of Pest Science has recently published a special issue on this pest, which was edited by Biondi and Desneux [5]. The special issue gave more consolidated and updated information on the moth biology, population dynamics, chemical and trophic ecology, and novel control technologies. This chapter gives concise information on *T. absoluta* biology and bionomics, economic significance, geographical distribution, invasive potential, natural enemies, and available management options.

2. Origin, morphology, and taxonomic position

T. absoluta originated in the Peruvian Central highlands from where it spread to other areas of Peru and then to the rest of Latin American countries during the 1960s [3]. TLM is small moth with body length of 5–7 mm and wingspan of 10–14 mm [8]. The moth has silvery-gray scales and black spots on the forewings. The antennae are long, filiform with black and brown scales (**Figure 1**). Shashank et al. [9] described the male and female genitalia as well as the pupal genital aperture as useful distinguishing character for sexing of the moth. Egg is small (0.36 mm long and 0.22 mm wide) with elliptical shape and creamy white to bright yellow color. Larva is whitish in first instar (0.9 mm long) and becomes greenish or light pink in the second and fourth instar (7.5 mm). Pupa is obtect with greenish coloration at first, turning to chestnut brown and dark brown near adult emergence [8] (**Figure 2**). Tabuloc et al. [10] studies the genome of *T. absoluta* to generate and design a panel of 21 SNP markers for the species identification instead of depending only on morphological identification and symptoms of damage on the host plants.

Tuta absoluta was originally described as *Phthorimaea absoluta* (Meyrick, 1917) in Peruvian Andes. The genus was changed to *Gnorimoschema* [11] and then to Scrobipalpula [12] and Scrobipalpuloides [13]. Povolny [14] corrected the currently

Figure 1.
Adult moth of tomato leafminer, Tuta absoluta (photo: Antonio Biondi).

Figure 2.
Larva (top) and pupa (bottom) of Tuta absoluta (photo: Antonio Biondi).

used name *Tuta absoluta*. The EPPO code and phytosanitary categorization for *T. absoluta* are GNORAB and EPP A1 action list no. 321, respectively [8].

3. Biology and bionomics

The TLM has a complete metamorphosis type of reproduction, where it undergoes through four developmental stages, namely, egg, larva, pupa, and adult (**Figure 3**). Adults are nocturnal and hide between host leaves during the day. The female starts to release a sex pheromone 1–2 days after emergence to lure males for mating. The female sex pheromone is a mixture of tetradecatrienyl acetate and tetradecadienyl acetate in a ratio of 10:1, respectively [15, 16]. TLM is known to have multiple mating and the average number of mating per female is about 10.4. Both sexes are polygamous with no refractory period. The female sometimes can exhibit deuterotoky parthenogenesis, which gives both females and males from unfertilized eggs [17]. Males use female sex pheromone to locate females and mating can last from few minutes to 6 hours. Female uses plant volatiles (kairomones) and leaf contact for oviposition. A single female can lay as many as 260 eggs during its life cycle, which may extend to 3 months [18]. About 92% of the total eggs are laid in the 1–3 days following mating [8]. Eggs are laid singly on the upper part of the plant (young leaves, stems, and sepals). The eggs hatch in 5–7 days depending on temperature and relative humidity. After hatching, the larvae go through four instars, which are completed in about 20 days. The mature larva then gets rid of all gut materials, constructs a silken cocoon, and turns into pre-pupa and pupa. Pupation may last for 10–11 days before adult emergence for female and male, respectively. Mature larvae leave the mines and build silken cocoon on the leaflet or in the soil. When pupation occurs in the mines or tomato fruit, the pre-pupa does not build cocoon. Adult longevity may extend for 30–40 days [8]. The whole life cycle of the moth is completed in 29–38 days, depending on the environmental conditions (**Figure 3**). Moreover, about 10–12 generation

Figure 3.
The life cycle of tomato leafminer, Tuta absoluta (photos: Antonio Biondi; design: Hamadttu A.F. El-Shafie).

may be produced annually. The thermal constant from egg to adult has been estimated to be 453.6 degree days (DD) [19]. TLM larvae do not enter diapause as long as food is available; however, it may overwinter as eggs, pupae, and adults [8, 18].

4. Host range

TLM is an oligophagous feeding on many related species of the family Solanaceae including tomato (*Solanum lycopersicum* L.), potato (*Solanum tuberosum* L.), eggplant (*Solanum melongena* L.), pepper (*Capsicum annuum* L.), sweet pepino (*Solanum muricatum* L.), tobacco (*Nicotiana tabacum* L.), the jimson weed (*Datura stramonium* L.), the African eggplant (*Solanum aethiopicum* L.), and the European black nightshade (*Solanum nigrum* L.) [1, 19, 20]. Sylla et al. [21] reported 12 host plants in the family Solanaceae, 2 in the Amaranthaceae, 2 in the Convolvulaceae, and 1 in the e in native South America, invaded Europe, and Africa. The two hosts in the Amaranthaceae are *Chenopodium album* L. and sugar beet, *Beta vulgaris* L., while the Fabaceae is represented by common bean, *Phaseolus vulgaris*.

TLM prefers tomato on which it is considered as a major pest while it is a minor pest on other alternative hosts. Host plant knowledge is essential for developing integrated pest management (IPM) against *T. absoluta* [21]. Sylla et al. [21] studied the oviposition acceptance, oviposition preference, and performance of two population of TLM from France and Senegal on six solanaceous plants, namely, tomato, eggplant, Ethiopian eggplant, potato, sweet pepper, and pepper. Their findings suggest that there is differentiation in the host range of TLM across invaded areas. In this respect, it has been reported that the relation between the female preference (maternity care) and larval performance should be very tight, as the larvae can survive only on small number of host plants [21].

5. Damage and economic significance

TLM usually attacks the apical buds, flowers, and new fruits of tomato. Larvae make conspicuous mines and galleries on leaves and stems. Damage can occur at any stage of tomato growth from seedlings to mature plant [8]. The larvae feed on the mesophyll tissue, leaving the epidermis intact, thus creating irregular mines and galleries on the leaves (**Figure 4**). The mines and galleries may become necrotic with time. This mining activities lead to reduction of the photosynthetic potential of infested leaves [1]. Infested tomato with TLM show burnt up-like symptoms [9]. The galleries made by the larvae are wider than that caused by the dipteran leaf miner *Liriomyza trifolii* [9, 22]. Larvae can penetrate the axillary buds of young stems when at high density. Thus, it leads to plant withering and check of vegetative growth [8].

After fruit setting, the larvae excavate tunnels in the fruits, which may facilitate invasion by pathogenic agents, resulting in fruit rot (**Figure 5**). The larvae of TLM have a cryptic behavior and endophagous habit, which makes detection of infestation at an early stage difficult [1]. Damage on stems causes necrosis that reduces tomato plant growth and development. Feeding tunnels and holes in the fruits lower their quality and reduce their market value [1]. The serious damage on tomato, due to *T. absoluta*, is caused by the leaf-mining activities and to a lesser extent by tunneling in the fruits [5]. Damage on tomato can reach 100% if no action is taken against the moth. Estimation of economic losses is difficult due to the interaction of many factors including climate, production pattern (greenhouse versus open field), and production costs including seeds, insecticides, fertilization, and other resources. Most of the damage occurs at the early years of invasion, due to lack of farmers' experience on how to manage the pest [1]. Han et al. [23] and Biondi and Desneux [5] summarized the damage of TLM into the following:

1. Production reduction due to injuries on leaves, stems, and fruits

2. Increase in cost of management practices (IPM) against the pest, particularly the purchase and application of insecticides

3. The ban or restriction of fresh tomato, from the side of non-invaded countries, which will affect the economy of countries where TLM is an endemic pest

4. Other costs include the disruption of preexisting integrated pest management (IPM) programs and disturbance of natural ecosystems [24].

Figure 4.
Symptoms of damage appear as mines and galleries on tomato leaves caused by feeding of T. absoluta larvae (photos: Antonio Biondi).

Figure 5.
Tunnels in ripe tomato fruits excavated by the larvae of Tuta absoluta (photos: Antonio Biondi).

6. Invasive potential and global distribution

According to Begon et al. [25], any species distribution is limited and governed by three basic components:

1. The ability of the species to reach a potential site (introduction pathway)

2. Capacity to develop in specific environmental conditions (establishment)

3. The ability to compete with other species occupying the same habitat

TLM is a highly invasive insect pest of tomato crop [1, 6]. The moth was first reported in Europe (Eastern Spain) in 2006 [19]. The introduction in Spain is believed to be from a single population in Chile [26]. Three years later, it was reported in Turkey, the fourth largest producer of tomato in the world, in 2009 [27]. It spread then across Europe and North African countries [6, 28] and Asian countries [23]. According to Seebens et al. [29], most of the invasion occurred during the last 40 years due to increased globalization and trading among continents. The possible introduction pathways for *T. absoluta* include tomato fruits, packing materials of tomato, eggplant and pepper, and planting material [30]. Santana et al. [31] studied the global geographic distribution of TLM using a combination of spatial distribution models as well as the current distribution of the pest. They showed that the suitable areas for *T. absoluta* include North and Central Americas, Africa, Europe, Asia, and Oceania at present time and in the future. Additionally, their model showed that large tomato-producing countries such as China, USA, and Mexico, where the moth is not present, stand a high risk of being invaded by *T. absoluta*. Damme et al. [32] and Han et al. [23] listed important reasons explaining the vast and wide spread of *T. absoluta* around the globe. These reasons include the following:

1. The strong intrinsic invasiveness with high reproductive potential of the moth

2. The dispersal capacity and ability of TLM to adapt to the newly invaded areas. The adults can fly actively for several kilometers, which allows for short distance spread [33].

3. The multivoltine reproductive cycle coupled with high overwintering capacity in greenhouses

4. The strong heat tolerance in open fields

5. Ability of the moth to develop on relatively large number solanaceous and non-solanaceous alternative hosts

The abovementioned reasons are pertaining to the biological traits of the pest. However, there are several reasons connecting human activities and measurements adopted by countries to curb the introduction, establishment, and spread of the pest that also contributed to the vast spread of *T. absoluta* [1, 4, 23, 34]. These reasons are as follows:

1. Weak and ineffective quarantine measures

2. Poor surveillance and phytosanitary measures

3. Bulk trade of untreated fresh tomato products

4. Rapid increase in the size of international trade and transportation of goods

5. Accidental transport of adults and other life cycle stages in consignments through containers and vehicles

6. Lack of joint efforts among affected countries in prevention and containment of the pest

7. Management

Because of its high invasiveness and economic significance, management of *T. absoluta* could be carried out at local, regional, and international levels. The management can be divided into pre-invasion and post-invasion measures. The former are mainly preventive including strict quarantine measures, inspection of tomato consignments, and treatment, when necessary, with proper fumigants before shipping. Endorsement, by countries, of *T. absoluta* as pest of a high risk in quarantine list is essential. *T. absoluta* population management in invaded countries could significantly lower the invasion risk to neighboring non-invaded ones [23, 35]. Establishment of regional network to connect research entomologists, policy-makers, and major stakeholders from all invaded as well as threatened countries [23]. Such network and platforms are supposed to coordinate joint research activities and validation of newly developed management technologies before being applied in the field.

The post-invasion management of *T. absoluta* is to try to eradicate the pest at an early stage of invasion if possible, otherwise a sustainable containment strategy based on integrated pest management is recommended.

In native and invaded areas in the world, current IPM components against TLM include the following:

1. Preventive measures and agronomic control

2. Semiochemically based control using female sex pheromone

3. Biological control

4. Biotechnological control

5. Chemical control by using selective insecticides

7.1 Preventive measures and agronomic control

Preventive and agronomic measures against TLM [36, 37] may include the following:

1. Destruction of previous crop remains to prevent the carryover of the pest

2. Removal of alternative hosts, particularly weeds from the genera *Solanum*, *Datura*, and *Nicotiana*

3. Exclusion of greenhouses with moth-proof sealing

4. Use pest-free planting material (transplants)

5. Screening of existing resistant tomato cultivars

6. Breeding transgenic resistant cultivars

7. Manipulation of soil trait and application of biofertilizers to enhance tomato plant resistance through bottom-up effects

8. Soil cultivation or covering with plastic mulch

7.2 Semiochemically based control

The female sex pheromone can be used in several ways for the management of TLM. These include the following:

1. Monitoring and surveillance. Pheromone-baited sticky traps can be used to monitor all stage of tomato production and across the production chain in nurseries, farms, greenhouses, and packaging and processing facilities [36]. Monitoring of TLM is performed by trapping males and/or by sampling eggs and larvae on infested tomato plants. The latter is however, tedious and difficult to perform over large areas. On the other hand, economic threshold based on male capture is not reliable because trapping process may be affected by

many factors such as population density of the moth, trap designs, and type of pheromone used. Light traps and water traps can also be used to capture both sexes.

2. Male annihilation by mass trapping of adults with pheromone traps (delta traps), which are usually efficient against newly introduced pest when population density is low. For mass trapping, it is recommended to use 20–40 traps/ha. A threshold of 3–4 moths per trap per week need to be reached before the beginning of mass trapping [36].

3. Mating disruption by saturating the atmosphere with sex pheromone, which alters ability of males to locate and find females. This technique can be effectively applied in confined environment such as protected tomato in greenhouses. However, the performance of the technique was poor [17].

7.3 Biological control

Salas Gervassio et al. [38] critically reviewed the natural enemies' complex in tomato agroecosystem. They determined the natural enemies that are suitable for augmentative and conservative strategies in South America and for classical biocontrol agents elsewhere in the world where *T. absoluta* has arrived. The authors reported that more than 50 species and morphospecies of Hymenoptera were associated with *T. absoluta*; however, only about 23 of them could be confirmed as parasitizing the moth. Augmentative biocontrol for *T. absoluta* is commercially available in South America using the parasitoid *Trichogramma pretiosum*, particularly in Brazil, Chile, Colombia, Ecuador, and Peru [38]. The use of endogenous natural enemies for biocontrol of TPW is one of the key points of conservative strategies [39]. The *Macrolophus basicornis* (Stal) and *M. pygmaeus* (Hemiptera: Miridae) are potential biocontrol agents against (egg predators) TLM. The nymphal stage of the former can consume an average of 331 eggs per day, while the adult can feed upon as many as 100 eggs per day [40, 41].

The parasitoids *Necremnus tutae* and *N. cosmopterix* (Hymenoptera: Eulophidae) are potential biocontrol agents against TLM [42].

The predator *Nesidiocoris tenuis* (Hemiptera: Miridae) can be used for the management of other tomato pests including the whiteflies, thrips, leafminers, and aphids [43]. This predator has shown great potential in controlling TPW in Asia [23], Turkey [44], and India [45]. This predator is commercially produced and released against TPW.

Omnivorous mirids had been used against TLM after its arrival in Europe through augmentative and inoculative release in the field and plant nurseries. They are sometimes supplied by conservation strategies using banker plants [1]. The mirid predators *Dicyphus bolivari* Lindberg and *D. errans* (Wolff) Hemiptera: Miridae are potential biocontrol agents against TPW [46].

The generalist egg parasitoid, *Trichogramma achaeae*, is a potential agent for biological control of *T. absoluta*. This worldwide-distributed parasitoid is also attracted by volatiles produced by tomato plants whether uninfested or infested as well as by the sex pheromone of the moth [47]. *Trichogramma evanescens* (Westwood) was also used against TLM in Turkey [44]. The egg parasitoid, *Trichogramma brassicae*, is a potential biocontrol agent of TPW [48]. Hemipteran predators such as anthocorids, geocorids, mirids, nabids, and pentatomids have been identified to be biological agents against *Tuta absoluta* [49]. Since larvae of *Tuta absoluta* are endophagous, cryptically living and feeding inside mines or tunnels in tomato leaves and fruit, respectively, their predation and parasitism by the natural enemies seem to be difficult. Nevertheless, numerous natural enemies can still be used in the management of this notorious pest. The eggs seem to be more vulnerable to predations and parasitism because they are exposed on the surface of tomato growing points. However, the efficacy of natural enemies in suppressing *T. absoluta* populations may be altered by environmental abiotic factors through bottom-up effects triggered by agronomic practices such as irrigation and fertilization. Moreover, plant constitutive and/or induced resistance traits against *T. absoluta* are another source of bottom-up effects, which may interact with irrigation and fertilization and jointly affect the performance and population density of *T. absoluta*, and counterpart natural enemies and their interactions [37]. In addition to the arthropod biocontrol agents, microbial biocontrol agents such as entomopathogenic nematode (EPN) of the genera *Steinernema* and *Heterohabditis* have potential to kill larvae of TLM when they are outside their mines.

7.4 Biotechnological control

Recently, the transcriptome data showed that most of the core genes of RNAi pathway such as Dicer-like and Argonaute and putative orthologous Sid-1 genes are present in *T. absoluta*, suggesting the feasibility of RNAi for controlling this pest [50]. Full plant protection and high larval mortality of *T. absoluta* have not been achieved, probably due to a low expression of dsRNA in transgenic plants [51]. Novel management technologies for TPW include genetically modified crops (GM), for example, GM *Bacillus thuringiensis* (Bt) tomato [52]. RNA interference (RNAi) is a biological mechanism that leads to posttranscriptional gene silencing directed by the presence of double-stranded RNA (dsRNA) molecules [53]. Biotechnically, sterile insect technique (SIT) may also be used for the management of TLM [37]. However, this technique may be compromised if field populations of *T. absoluta* can reproduce by deuterotokous parthenogenesis [17]. It is worth to mention here that these authors reported tychoparthenogensis reproduction of *T. absoluta* under laboratory conditions. They stated that the origin of this type of reproduction could be considered as classical automictic tychoparthenogensis or due to the microbial manipulation by bacterial endosymbiont such as *Wolbachia*, which has recently been identified in *T. absoluta* [54].

7.5 Chemical control

Chemical control of the invasive TLM is difficult; however, its arrival to new invaded areas has been linked to an excessive application of broad-spectrum insecticides [1, 6, 55], in attempts to curb the outbreaks of the pest and to reduce yield losses in tomato crop. Currently, insecticides application seems to be the most commonly used strategy against *T. absoluta* worldwide in open fields of tomato [1, 56–58]. The cryptic behavior and the endophagous habit of larvae make it extremely difficult to control TLM with insecticides [1, 19]. The possible reasons for difficulty of controlling TLM with insecticides, according to Biondi et al. [1] and Guedes et al. [58] include the following:

1. Infestation of tomato by the moth occurs at an early stage of plant growth

2. The multiple attacks by the pest on different plant parts (stems, leaves, buds, young fruit, and ripe fruit)

3. The morphology and architecture of tomato plant that provide protection for feeding larvae against insecticides

Insecticides from different chemical classes were used against TLM in South America, Europe, and other parts of the world. These chemical classes include, but not limited to, organophosphate, pyrethroids, pyrrole, spinosyns, diamides, benzoylureas, and avermactins [54, 56, 59]. Spinosad, azadirachtin, and *Bacillus thuringiensis* toxins (Bt) have been to control TLM in organic tomato production systems [1, 56].

The excessive application of insecticides to prevent and control the outbreaks of *T. absoluta*, particularly in open fields lead to an increased selection pressure which, eventually reduce the effectiveness of such insecticides [58, 60]. For example, when the moth was introduced in Brazil, the farmers initially used insecticides at frequencies of 10–12 applications per cropping season, which was later increased to 30 applications [60]. In Turkey, the annual cost of chemical insecticides used against *T. absoluta* in 2014 was about 160 million Euros [27]. The frequent use of insecticides speeded the appearance of resistance in tomato leaf miner populations, which can migrate outside their geographical range into new invaded areas [1, 6, 56–59].

Guedes et al. [58] reported that enhanced levels of detoxification enzymes and altered target sites are the main resistance mechanisms commonly found in *T. absoluta*. In addition to the development of resistance in TLM populations, due to excessive use of insecticides, compromising of biological control, in tomato agroecosystems, is also not avoidable. In this respect, Soares et al. [43] studied the lethal and sublethal effects of five insecticides (spinetoram, chlorantraniliprole + abamectin, triflumuron, tebufenozide, and abamectin) on adults and the third instar nymph of the predator *Macrolophus basicornis*. They concluded that abamectin caused high mortality in both adult and nymphs. All tested insecticides caused negative effect on the predator.

To overcome the problems of insecticide resistance and other harmful effects on tomato ecosystem, due to the excessive use of insecticides, insecticide resistance management (IRM) strategies are needed to sustain production of tomato crop [1, 58]. Such strategies include adoption of alternative control options such as cultural control, semiochemically based control, biological control, and host plant resistance. All these alternative strategies and tactics would reduce the reliance on insecticides and accordingly the selection pressure on TLM populations [1, 58].

8. Conclusions

Recently, tomato leaf miner has emerged as a highly invasive key pest threatening the global production of tomato. The global commercialization and trade of fresh tomato fruit and transplanting material have accelerated the spread of the pest. The impact of *T. absoluta* on global tomato production industries and on the livelihood of small tomato farming communities in Africa and Asia might be more severe in the coming years, unless great efforts are made to contain its spread. Chemical control of *T. absoluta* with insecticides seems to be ineffective and not sustainable; therefore, alternative management options such as biological control and semiochemically based control should be encouraged. The socioeconomic impact of this moth on subsistent agriculture need to be addressed in future studies.

Author details

Hamadttu Abdel Farag El-Shafie
Date Palm Research Center of Excellence, King Faisal University, Al-Ahsa, Saudi Arabia

*Address all correspondence to: elshafie62@yahoo.com

References

[1] Biondi A, Guedes RNC, Wan FH, Desneux N. Ecology, worldwide spread and management of the invasive South American tomato pinworm, *Tuta absoluta*: Past, present and future. Annual Review of Entomology. 2018;**63**:239-258

[2] Cameron EK, Vilá M, Cabezan M. Global meta-analysis of the impacts of terrestrial invertebrate invaders on species, communities and ecosystems. Global Ecology and Biogeography. 2016;**25**:596-606

[3] Campos MR, Biondi A, Adiga A, Guedes RNC, Desneux N. From the Western Palaearctic region to beyond: *Tuta absoluta* 10 years after invading Europe. Journal of Pest Science. 2017;**90**:787-796

[4] Bradshaw CJ, Leroy B, Bellard C, Roiz D, et al. Massive yet grossly underestimated global costs of invasive insects. Nature Communications. 2016;**7**:12986

[5] Biondi A, Desneux N. Special issue on *Tuta absoluta* recent advances in management methods against the background of an ongoing worldwide invasion. Journal of Pest Science. 2019;**92**:1313-1315

[6] Desneux N, Luna MG, Guillemaud T, Urbaneja A. The invasive South American tomato pinworm, *Tuta absoluta*, continues to spread in Afro-Eurasia and beyond: The new threat to tomato world production. Journal of Pest Science. 2011;**84**:403-408

[7] Olaniyi JO, Akanbi WB, Adejumo TA, et al. Growth, fruit yield, and nutritional quality of tomato varieties. African Journal of Food Science. 2010;**4**:398-402

[8] EPPO. Data sheet on quarantine pests: *Tuta absoluta*. EPPO Bulletin. 2005;**35**:434-435

[9] Shashank PR, Chandrashekar K, Meshram NM, Sreedevi K. Occurrence of *Tuta absoluta* (Lepidoptera: Gelechiidae) an invasive pest from India. Indian Journal of Entomology. 2015;**77**(4):323-329

[10] Tabuloc CA, Lewald KM, Conner WR, et al. Sequencing of *Tuta absoluta* genome to develop SNP genotyping assays for species identification. Journal of Pest Science. 2019;**92**:1397-1407

[11] Clarke JF. New species of microlepidoptera from Japan. Entomological News. 1962;**73**:102

[12] Povolny D. Gnorimoschemini Trib. Nov. Eine Neue Tribus der Familie Gelechiidae nebst Bemerkungen zu ihere Taxonomie (Lepidoptera). Acta societatis Entomologicae Cechosloveniae. 1964;**61**(4):330-359

[13] Povolny D. Gnorimoschemini of South America III: The Scrobipaluloid genera (Insecta, Lepidoptera, Gelechiidae). Steepnstrupia. 1987;**13**:1-91

[14] Povolny D. Gnorimoschemini of South America VI: Identification keys checklist of neotropical taxa and general consideration (Insecta, Lepidoptera, and Gelechiidae). Steepnstrupia. 1994;**20**:1-42

[15] Attygalle AB, Jham GN, Svatos A, Frighetto RTS, Meinwald J, Vilela EF, et al. Microscale, random reduction: Application of the characterization of (3E, 8Z, 11Z)-3, 8, 11-tetradecatrienyl acetate, a new lepidopteran sex pheromone. Tetrahedron Letters. 1995;**36**(31):5471-5474

[16] Attygalle AB, Jham GN, Svatos A, Frighetto RST, Ferrara FA, Vilela EF, et al. (3E, 8Z, 11Z)-3, 8, 11-tetradecatrienyl acetate, major

sex pheromone component of the tomato pest *Scrobipalpuloides absoluta* (Lepidoptera: Gelechiidae). Bioorganic & Medicinal Chemistry. 1996;**4**(3):305-314

[17] Caparros Megido R, Haubruge E, Verheggen FJ. First evidence of deuterotokous parthenogenesis in the tomato leafminer, *Tuta absoluta* (Meyrick) (Lepidoptera: Gelechiidae). Journal of Pest Science. 2012;**85**:409-412

[18] Uchoa-Fernandes MA, Della Lucia TMC, Vilela EF. Mating, oviposition and pupation of *Scrobipalpuloides absoluta* (Meyr.) (Lepidoptera: Gelechiidae). Anais da Sociedade Entomologica do Brasil. 1995;**24**:159-164

[19] Desneux N, Wajnberg E, Wyckhuys KAG, Burgio G, et al. Biological invasion of European tomato crops by *Tuta absoluta*: Ecology, geographic expansion and prospects for biological control. Journal of Pest Science. 2010;**83**:197-215

[20] EPPO. EPPO Plant Quarantine Data Retrieval System PQR Version 4.6. 2007. Available from: http://www.eppo.org/DATABASES/pqr/pqr.htm

[21] Sylla S, Brevault T, Monticell LS, Diarra K, Desneux N. Geographic variation of host preference by the invasive tomato leaf miner *Tuta absoluta*: Implication for host range expansion. Journal of Pest Science. 2019;**92**:1387-1396

[22] Nayana BP, Kalleshwaraswamy CM. Biological and external morphology of invasive tomato leaf miner, *Tuta absoluta* (Meyrick) (Lepidoptera: Gelechiidae). Pest Management in Horticultural Ecosystems. 2015;**21**(2): 169-174

[23] Han P, Bayram Y, Shaltiel-Harpaz L, et al. *Tuta absoluta* continues to disperse in Asia: Damage, ongoing management and future challenges. Journal of Pest Science. 2019;**92**:1317-1327

[24] Paini DR, Sheppard AW, Cook DC, De Barro PJ, Worner SP, Thomas MB. Global threat to agriculture from invasive species. Proceedings of the National Academy of Sciences of the United States of America. 2016;**113**:7575-7579

[25] Begon M, Townsend CR, Harper JL. Ecology: From Individuals to Ecosystem. 5th ed. Malden: Blackwell Publishing; 2005

[26] Guillemaud T, Blin A, Le Goff I, et al. The tomato borer, *Tuta absoluta*, invading the Mediterranean Basin, originates from a single introduction from Central Chile. Science Reports. 2015;**5**:8371

[27] Oztemiz S. *Tuta absoluta* Povolny (Lepidoptera: Gelechiidae), the exotic pest in Turkey. Roman Journal of Biology and Zoology. 2014;**59**:47-58

[28] Mansour R, Brévault T, Chailleux A, Cherif A, Grissa-Lebdi K, et al. Occurrence, biology, natural enemies and management *Tuta absoluta* in Africa. Entomologia Generalis. 2018;**38**:83-111

[29] Seebens H, Blackburn TM, Dyer EE, Gerovesi P, et al. No saturation in the accumulation of alien species worldwide. Nature Communications. 2017;**8**:14435

[30] Karadjova O, Iliva Z, Krumov V, Petrova E, Ventsislavov V. *Tuta absoluta* (Meyrick) (Lepidoptera: Gelechiidae): Potential for early entry, establishment and spread in Bulgaria. Bulgarian Journal of Agricultural Science. 2013;**19**(3):563-571

[31] Santana PA Jr, Kumar L, Da Silva RS, Picanco MC. Global geographic distribution of *T. absoluta* as affected by climate change. Journal of Pest Science. 2019;**92**:1373-1385

[32] Damme VV, Berkvens N, Moerkens R, et al. Overwintering potential of the invasive leafminer *Tuta absoluta* (Meyrick) (Lepidoptera: Gelechiidae) as a pest in greenhouse tomato in Western Europe. Journal of Pest Science. 2015;**88**:533-541

[33] Stol W, Griepnik FC, van Deventer P. *Tuta absoluta* a new pest for tomato production in Europe. II Jornades Feromonas Murcia. 2009. Available from: http://www.feromonasmurcia.es/ingles/files/textosing/hortalizas/H-1%20W%20STOL%20TUTA%20IN.pdf

[34] Potting R. Pest Risk Analysis, *Tuta absoluta*, Tomato Leaf Miner Moth. Utrecht, The Netherlands: Plant Protection Service of the Netherlands; 2009. p. 24

[35] Xian XQ, Han P, Wang S, Zhang G, et al. The potential invasion risk and preventive measures against the tomato leafminer *Tuta absoluta* in China. Entomologia Generalis. 2017;**36**:319-333

[36] IRAC. *Tuta absoluta* - The Tomato Leaf Miner or Tomato Borer. Recommendations for Sustainable and Effective Resistance Management. 2014. Available from: http://www.irac-online.org/pests/tuta-absoluta/presentations/ [Accessed: May 2020]

[37] Han P, Desneux N, Becker C, Larbat R, et al. Bottom-up effects of irrigation, fertilization and plant resistance on *Tuta absoluta*: Implications for integrated pest management. Journal of Pest Science. 2019;**92**:1359-1370

[38] Salas Gervassio NG, Aquino D, Vallina C, Biondi A, Luna MG. A re-examination of *Tuta absoluta* parasitoids in South America for optimized biological control. Journal of Pest Science. 2019;**92**:1343-1357

[39] Naranjo S, Ellsworth PG, Frisvoid GB. Economic value of biological control in integrated pest management of managed plant systems. Annual Review of Entomology. 2015;**60**:621-645

[40] Van Lenteren C, Bueno J, Smith JVH, et al. Predation of *Tuta absoluta* eggs during the nymphal stages of three neotropical mirid predators on tomato. Bulletin of Insectology. 2017;**70**:69-74

[41] Soares MA, Campos MR, Passos LC, Carvalho GA, et al. Botanical insecticides and natural enemies: A potential combination for pest management against *Tuta absoluta*. Journal of Pest Science. 2019;**92**:1433-1443

[42] Bodino N, Ferracini C, Tavella L. Functional response and age-specific foraging behavior of *Necremnus tutae* and *N. cosmopterix*, native natural enemies of the invasive pest *Tuta absoluta* in Mediterranean area. Journal of Pest Science. 2019;**92**:1467-1478

[43] Soares MA, Passos LC, Campos MR, Collares LJ, Desneux N, Carvalho GA. Side effect of insecticides commonly used against *Tuta absoluta* on the predator *Macrolophus basicornis*. Journal of Pest Science. 2019;**92**:1447-1456

[44] Kececi M, Oztop A. Possibilities for biological control of *Tuta absoluta* (Meyrick, 1917) (Lepidoptera: Gelechiidae) in the western Mediterranean region of Turkey. Turkish Journal of Entomology. 2017;**41**:219-230

[45] Sankarganesh E, Firake DM, Sharma B, et al. Invasion of the South American tomato pinworm, *Tuta absoluta*, in northeastern India: A new challenge and biosecurity concerns. Entomologia Generalis. 2017;**36**:335-345

[46] Ingegno BL, Messelink GJ, Bodino N, et al. Functional response of the mirid predators *Dicyphus bolivari*

and *D. errans* and their efficacy as biological control agents of *T. absoluta* on tomato. Journal of Pest Science. 2019;**92**:1457-1466

[47] Gontijo L, Cascone P, Giorgini M, et al. Relative importance of host and plant semiochemicals in the foraging behavior of *Trichogramma achaeae*, an egg parasitoid of *Tuta absoluta*. Journal of Pest Science. 2019;**92**:1479-1488

[48] Mohamadi P, Razmjou J, Naseri B, Hassanpour M. Humic fertilizer and vermicompost applied to the soil can positively affect population growth parameters of *Trichogramma brassicae* (Hymenoptera: Trichogrammatidae) on eggs of *Tuta absoluta* (Lepidoptera: Gelechiidae). Neotropical Entomology. 2017;**46**:678-684

[49] Ferracini C, Bueno VHP, Dindo ML, Ingegno BL, et al. Natural enemies of *Tuta absoluta* in the Mediterranean basin. Biocontrol Science and Technology. 2019;**29**(6):578-609. DOI: 10.1080/09583157.2019.1572711

[50] Camargo RA, Herai RH, Santos LN, Bento FM, Lima JE, Marques-Souza H, et al. De novo transcriptome assembly and analysis to identify potential gene targets for RNAi-mediated control of the tomato leaf miner (*Tuta absoluta*). BMC Genomics. 2015;**16**:635

[51] Camargo RA, Barbosa GO, Possignolo IP, Peres LE, Lam E, Lima JE, et al. RNA interference as a gene-silencing tool to control *Tuta absoluta* in tomato (*Solanum lycopersicum*). PeeJ. 2016, 2016;**4**:e2673. DOI: 10.7717/peerj.2673 [Accessed: 28 May 2020]

[52] Selale H, Dagli F, Mutlu N, et al. Cry1Ac-mediated resistance to tomato leafminer (*Tuta absoluta*) in tomato. Plant Cell Tissue and Organ Culture. 2017;**131**:65-73

[53] Majidiani S, PourAbad RF, Laudani F, et al. RNAi in *Tuta absoluta* management, effects of injection and root delivery of dsRNAs. Journal of Pest Science. 2019;**92**:1409-1419

[54] Ŝkaljac M, Kostanjsek R, Zanic K. The presence of *Wolbachia* in *Tuta absoluta* (Lepidoptera: Gelechiidae) populations from coastal Croatia and Montenegro. African Entomology: Journal of the Entomological Society of Southern Africa. 2012;**20**(1):191-194

[55] Balzan MV, Moonen AC. Management strategies for the control of *Tuta absoluta* (Lepidoptera: Gelechiidae) damage in open-field cultivation of processing tomato in Tuscany (Italy). EPPO Bulletin. 2012;**42**:217-225

[56] Silva GA, Picanco MC, Bacci L, et al. Control failure likelihood and spatial dependence of insecticide resistance in the tomato pinworm, *Tuta absoluta*. Pest Management Science. 2011;**67**:913-920

[57] Guedes RNC, Picanco MC. The tomato borer, *Tuta absoluta* in South America: Pest status, management and insecticide resistance. EPPO Bulletin. 2012;**42**:211-216

[58] Guedes RNC, Roditakis E, Campos MR, et al. Insecticide resistance in the tomato pinworm, *Tuta absoluta* pattern, spread, mechanisms, management and outlook. Journal of Pest Science. 2019;**92**:1329-1342

[59] Silva JE, Ribeiro LMS, Vinasco N, Guedes RNC, Siqueira HAA. Field-evolved resistance to chlorantraniliprole in tomato pinworm *Tuta absoluta*. Inheritance, cross-resistance profile, and metabolism. Journal of Pest Science. 2019;**92**:1421-1431

[60] Guedes RNC, Siqueira HAA. The tomato borer *Tuta absoluta*: Insecticide resistance and control failure. CAB Reviews: Perspectives in Agriculture, Veterinary Science, Nutrition and Natural Resources. 2012;**7**:1-7

Section 2

Invasive Plant Species

Chapter 6

Hybridogenic Activity of Invasive Species of Asteraceae

Maria A. Galkina and Yulia K. Vinogradova

Abstract

We studied taxa from genus *Bidens*, *Solidago*, and *Erigeron*, sect. *Conyza* (Asteraceae). By analyzing the nucleotide sequences of the internal transcribed spacer (ITS1)-ITS2 site, the hybrid origin of the *Bidens* × *decipiens*, previously attributed to the North American alien species *B. connata*, was confirmed. The analysis of trnL-trnF sequences showed that the native *B. cernua* is the maternal species and the invasive *B. frondosa* is the paternal species in all probabilities. Diagnostic morphological features of the three *Solidago* taxa growing together in the vicinity of Pskov have been studied: a native *S. virgaurea*, an invasive species of North American origin *S. canadensis*, and their hybrid *S.* × *niederederi*. *S.* × *niederederi* has an intermediate position between *S. virgaurea* and *S. canadensis*. The hybrid origin of *S.* × *niederederi* is proven by molecular analysis of nuclear DNA nucleotide sequences (ITS1-ITS2 site). It is not yet possible to unambiguously answer the question which parent species is maternal and which is paternal. We also studied invasive species of the genus *Erigeron* sect. *Conyza* in the Mediterranean. Occasionally occurring in Southern Europe, individuals of *E. canadensis* × *E. sumatrensis* with intermediate morphological features, described as "*Conyza* × *rouyana*," are likely unstable and soon "absorbed" by the parent species *E. sumatrensis*. Contrary to the hypothesis by C. Elton explaining the success of plant invasion in a new homeland by strengthening hybridization processes in the secondary distribution range.

Keywords: hybrids, hybridization, invasive species, *Bidens* × *decipiens*, *Solidago* × *niederederi*, *Erigeron canadensis*, *E. sumatrensis*, ITS1-ITS2 site, *rpl32-trnL* intergenic spacer, *trnL-trnF* intergenic spacer

1. Introduction

There is a hypothesis that the strengthening of hybridization processes in the secondary distribution range contributes to more successful existence of plants in their new homeland [1, 2]. Under unusual conditions, alien species can form hybrids with closely related native species, as well as with other alien plants inhabiting a given area. Often, hybrids are better adapted to secondary distribution range conditions than parent taxa [3–5], resulting in landscapes in a new home area. Successful recombination of genetic traits of parent species reduces the lag phase (a period of adaptation of an alien taxon to new conditions during which there is not yet active introduction into natural phytocoenosis and expansion of the secondary distribution range) and leads to the formation of new active "species transformers."

Cross-pollinated plant species are the most predisposed to hybridization. Sympatric species are less likely to be cross-species hybridized than allopatric species or populations [6]. The share of hybrid taxa among invasive species of Middle Russia reaches 10% [7].

For a long time, the most important factor in limiting hybridization was geographical isolation, but nowadays closely related taxa come into contact with each other through a multitude of anthropogenic "corridors" [8]. Thus, inter-regional immigration occurs by means of the introduction of plants, which can be frequent and repetitive, and therefore it significantly increases immigration flow [9]. If we consider the situation where conditions for the hybridization of closely related taxa already exist, there may be several possible developments that coexist: (a) New hybrid taxon may appear, and (b) native taxon disappears. During hybridization, genetic assimilation occurs, and new genes are injected into one or both parent species. Hybrids, even being fertile, can, however, be reproductively isolated from parent plants due to the effect of the selection on reproductive traits (allopolyploidy, heterozygous translocations, recombination, mitochondrial DNA-specific differences) and/or due to factors that predetermine crossing (flowering phenology, separation of ecological niches). Interspecific hybridization may also facilitate the naturalization of rare genotypes and cause an increase in their proportion by inverse crossbreeding with alien parent taxa or hybridization between the hybrids themselves. Greater selection advantages for alien alleles should lead to faster replacement of natural alleles through hybridization and slower replacement without hybridization [10]. The period of displacement (substitution) decreases significantly with increasing immigration flow and selective differentiation. Immigration and selection operate in a variety of ways: increasing immigration levels result in the substitution of native species by suppressing them, while increasing selective differentiation in favor of an alien species results in the substitution of an alien species by genetic assimilation without leaving "pure" native species. At moderate and high immigration rates, the loss of native species can be rapid with or without hybridization. Given the high number of species introduced by humans, the loss of native species can increase only as a result of hybridization [11].

Hybridization increases the threat of extinction of many species due to introgression [12, 13]. High degree of introgression is often manifested by wind-pollinated species such as oaks. Hybridization and introgression can lead to a hybrid complex consisting of many hybrids due to a large number of loci. Thus, multiloci seem to increase the number of hybrid types and genetic complexes and accelerate the reduction of "pure" natural species. In addition, the large number of loci essentially reduces the probability of having a "pure" individual of any parent origin [11]. Without introgression, hybrids, being reproductively isolated, can quickly form a new species. With introgression, speciation slows down as inverse crosses with parent lines occur. The impact of hybridization and introgression on the rate of substitution of native species by closely related ones has been addressed by a mathematical model involving a one-loop bipartite inheritance scheme with different levels of cross-species hybridization [11]. Although the model did not take into account vegetatively propagating hybrids, the results showed that the substitution of natural taxa by alien ones could occur very quickly (in less than five generations). According to the results, hybridization and introgression can increase the degree of substitution of native species by non-native ones. Introgression increases species substitution with low immigration, but prevents substitution when an indigenous species has a significant advantage in selection as well as with higher immigration levels. However, as introductions are associated with increased frequency of hybrids, the impact on the indigenous taxon remains high, and the likelihood of extinction increases significantly [11].

It is known that the highest invasive activity is exhibited by species of Asteraceae family [14], so we have focused our attention on representatives of this group of alien species. It is often impossible to say with certainty whether plants with intermediate morphological features are hybrids (between two species of the same genus). This may also be the case for new ecological forms, resulting from microevolution of species. To confirm or disprove the hypothesis about hybrid origin of certain taxon, it is best to use molecular genetic methods.

2. Materials and methods

DNA was extracted from silica gel dried leaves of *Bidens*, *Solidago*, and *Erigeron* taxa according to the method by Rogers and Bendich [15]. The herbarium specimens are stored in the herbarium of the Tsitsin Main Botanical Garden (MHA). Polymerase chain reaction (PCR) was carried out in a DNA Engine Dyad Peltier Thermal Cycler amplifier (Bio-Rad, United States). For the nuclear ribosomal internal transcribed spacer (ITS) 1-2 (ITS1-ITS2), *nnc18s10* (forward) and *c26A* (reverse) primers with an annealing temperature of 50°C were used. For the chloroplast loci (*rpl32-trnL* and *trnL-trnF* intergenic spacers), primers were used at the annealing temperature from 0.3 to 65°C [16]. For the chloroplast locus *rpl32-trnL*, we used primers *rpl32F* (forward) and *trnL* UAG (reverse), and for the other chloroplast locus *trnL-trnF*, we used primers *c* (forward) and *f* (reverse). Purification of the PCR product for sequencing was carried out in a mixture of ammonium acetate with ethanol. The nucleotide DNA sequences were determined on an automatic sequencer (Syntol). Further processing of the nucleotide sequences was carried out in the BioEdit program. The data were sent to GenBank (2019), in which these nucleotide sequences can be found by the additional numbers assigned to them (**Table 1**). Phylogenetic trees were constructed using SplitsTree4.

3. *Bidens* × *decipiens*

Bidens connata Muhl. ex Willd. is a North American species whose natural area extends from Alaska in the north to Mexico in the south [17]. The species has high polymorphism within native area, and several varieties have been described: *B. connata* var. *ambiversa* Fassett, var. *anomala* Farwell, var. *fallax* (Warnstorf) Sherff, var. *gracilipes* Fernald, var. *inundata* Fernald, var. *petiolata* (Nuttall) Farwell, var. *pinnata S. Watson*, and var. *submutica* Fassett [18, 19]. In the second half of the twentieth century, American botanists made suggestions about the hybrid nature of *B. connata* based on morphological features [20]. This species was indicated as an alien for Europe [21]. However, European plants called "*B. connata*" are morphologically different from American samples. Their outer leaves are clearly leaf-shaped, well-developed, and 3–6 cm long, with no reedy flowers, and the first real leaves are less narrow and with more pronounced petioles than those of *B. connata* and fewer denticles on the leaves, and the denticles, in turn, are usually larger and less regularly located [22]. Plants from European populations have been described as *B.* × *decipiens* Warnst. in 1895. The typical excrement material collected by Carl Warnstorf is stored in the herbaria of Edinburgh (E), Frankfurt (FR), and Charles University in Prague (PRC) [23]. In Europe, the locations of *B.* × *decipiens* are few and far between. A map of the gradual eastward expansion of this species was previously compiled by the authors of this paper [24] and is shown in **Figure 1**. Previously, we studied morphological features of *B.* × *decipiens* in Russia and found that features of this species are intermediate between the North American invasive

Sample no.	Number of ITS/rpl32–trnL/trnL–trnF sequence in GeneBank	Taxon	Date and place of collection, notes
de_1a	MK559763/ – /MK575566	*Bidens* × *decipiens* (= "*B. connata*")	Russia, Kaluga region, Milyatinsky Reservoir, 2013 54.4914° N, 34.3393° E
de _1b	MK559764/ – /MK575567		
de _1c	MK559765/ – / MK575568		
de _1d	MK559766/ – /MK575569		
de _2a	MK559767/ – /MK575570		Russia, Kaliningrad region, 2013 54.95° N, 20.49° E
de _2b	MK559768/ – /MK575571		
de _2c	MK559769/ – /MK575572		
de _2d	MK559770/ – / –		
de _3a	MK559771/ – /MK575573		Russia, Vladimir region, near Tasinsky village, 2014 Formed with dissected lower leaves 55.567° N, 40.172° E
de _3b	MK559772/ – / –		
de _3c	MK559773/ – / MK575574		
de _4a	MK559774/ – /MK575575		Russia, Vladimir region, near Tasinsky village, 2014 Formed with whole lower leaves 55.567° N, 40.172° E
de _4b	MK559775/ – /MK575576		
fr_5a	MK559780/ – /MK575581	*B. frondosa*	Russia, Vladimir region, near Tasinsky village, 2014 55.567° N, 40.172° E
fr_5b	MK559781/ – /MK575582		Russia, Vladimir region, near Tasinsky village, 2018 55.567° N, 40.172° E
fr_5c	MK559782/ – /MK575583		
cr_6a	MK559755/ – /MK575559	*B. cernua*	Russia, Moscow region, near Zvenigorod town, 2014 55.69° N, 36.74° E
cr_6b	MK559756/ – /MK575560		
t_7	MK559754/ – /MK575558	*B. tripartita*	Russia, Vladimir region, near Tasinsky village, 2018 55.567° N, 40.172° E
cr_8a	MK559757/ – /MK575561	*B. cernua*	
cr_8b	MK559758/ – /MK575562		
cr_8c	MK559759/ – /–		
cr_9a	MK559760/ – /MK575563		Belarus, Dzerzhinsk, 2018 53.693° N, 27.165° E
cr_9b	MK559761/ – /MK575564		
cr_9c	MK559762/ – /MK575565		
fr_10a	MK559783/ – /MK575584	*B. frondosa*	
fr_10b	MK559784/ – /MK575585		
de _11a	MK559776/ – /MK575577	*B. decipiens* (= "*B. connata*")	
de _11b	MK559777/ – /MK575578		
de _11c	MK559778/ – /MK575579		
de _13	MK559779/ – /MK575580		Russia, Moscow, park near Sviblovo estate, 2018 55.8639° N, 37.6396° E
v_1a	MK491849/MK474079/ –	*Solidago virgaurea*	Russia, Pskov region, Pskov district, vicinity of Pskov, idle field, 2018 57.80° N, 28.25° E
v_1b	MK491850/MK474080/ –		
v_1c	MK491851/MK474081/ –		
n_2a	MK491852/MK474082/ –	*S.* × *niederederi*	

Sample no.	Number of ITS/rpl32–trnL/trnL–trnF sequence in GeneBank	Taxon	Date and place of collection, notes
n_2b	MK491853/MK474083/ –		
n_2c	– /MK474084/ –		
c_3a	MK491854/MK474085/ –	*S. canadensis*	
c_3b	MK491855/MK474086/ –		
c_3c	MK491856/MK474087/ –		
v_4	– /MK474088/ –	*S. virgaurea*	Russia, Moscow region, Chekhov district, near the village of Chudinovo, idle field, 2018 55.1° N, 37.5° E
c_5a	MK491857/MK474090/ –	*S. canadensis*	
c_5b	– /MK474089/ –		
v_6a	MK491858/– / –	*S. virgaurea*	Russia, Moscow region, "Losiny Ostrov" National Park, Pine forest, 2014 55.89° N, 37.77° E
v_6b	MK491859/– / –		
3	MK397980/– / –	*Erigeron sumatrensis*	Italy, Pompeii, 2016 40.7° N, 14.5° E
5a	MK397981/– / –		Italy, the Island of Ischia, 2016 40.7° N, 13.9° E
5b	MK397982/– / –		
5c	MK397983/– / –		
6	MK397984/– / –		Italy, Herculaneum, 2016 40.8° N, 14.4° E
8a	MK397986/– / –	*E. sumatrensis* × *E. canadensis* (?)	Italy, Naples, 2016 40.8° N, 14.2° E
8b	MK397987/– / –		
10a	MK397988/– / –		Italy, Pompeii, 2016 40.7° N, 14.5° E
10b	MK397989/– / –		
13a	MK397991/– / –	*E. canadensis*	Italy, the Island of Ischia, 2016 40.7° N, 13.9° E
13b	MK397992/– / –		
16	MK397985/– / –		Portugal, Lisbon, 2017 38.7° N, 9.1° W
18	MK397993/– / –		Spain, Madrid, park, 2017 40.4° N, 3.7° W
19	MK397994/– / –	*Erigeron* sp.	
20	MK397995/– / –	*E. canadensis*	
22	MK397990/– / –		

Note: "E. sumatrensis × E. canadensis (?)" – putative hybrids.

Table 1.
Samples of the studied taxa of Asteraceae.

B. frondosa L. and the native *B. cernua* L. *B. × decipiens* are covered with two types of hairs—duplex, from two cells (as in *B. frondosa*), and simple multicellular (as in *B. cernua*). In addition, the seeds of *B. × decipiens* are quadrilateral and have four axes (as in *B. cernua*) and are covered with warts (as in *B. frondosa*). Heads of *B. × decipiens* are similar in size and shape to heads of *B. frondosa*, and the leaves are whole, as in *B. cernua*. On the basis of these data, we hypothesize the hybrid origin of *B. × decipiens* [25].

Nucleotide sequence analysis shows that not all individuals defined as *B. × decipiens* can be called hybrids. The point is that in cases of nucleotide substitutions

Figure 1.
Map of the secondary distribution range of Bidens × decipiens.

at the ITS1-ITS2 site differentiating *B. frondosa* and *B. cernua*, we encounter heterozygosity of samples of *B.* × *decipiens* (and, accordingly, ambiguity of reading of the sequence) in many cases, but still not in all (as it should be expected for the hybrid F). At the same time, each of our samples of *B.* × *decipiens* is characterized by at least some such ambiguous readings of nucleotides (C or T, A or T, A or C) (**Figure 2**) in the case of substitutions, which, of course, confirms the hypothesis of the hybrid origin of this taxon. Populations of *B.* × *decipiens* from different parts of the range differ in the number of substitutions. Thus, samples from the banks of the Milyatinskoye reservoir in the Kaluga region demonstrate in most cases the presence of ambiguous readings in the case of nucleotide substitutions, with the exception of sample de_1a, in which heterozygosity is not observed in all cases of substitutions. It means that this population is a hybrid one. In addition, both parent species grow on the banks of the Milyatinskoye reservoir in close proximity to the population of *B.* × *decipiens*, which indirectly supports this view [26]. Probably, the differences in the sample de_1a are due to the presence of introgression, i.e., this sample is a backcross resulting from crossing of *B.* × *decipiens* with *B. cernua*, because ITS1-ITS2 of this sample has a stronger DNA area similar to *B. cernua* than others. The same situation is observed with individuals of *B.* × *decipiens* from Belarus. *B.* × *decipiens* specimens from the Kaliningrad region, by contrast, in most cases show similarities with *B. cernua* in the case of substitutions rather than heterozygosity, except for sample de_2a. In this case, there are two possible variants—in the first case, we collected samples of backcrosses and see the result of introgression; in the second case, the parent form is another form of *B. frondosa*, not the widespread *B. frondosa* var. *frondosa*. The second variant is less probable. However, other forms of *B. frondosa* have been recently found [27]. Among plants of *B.* × *decipiens* collected in the Vladimir region, two forms distinct on lower leaves—with a dissected sheet plate (samples de_3a, de_3b, and de_3c) and with a whole sheet plate (de_4a, de_4b)—are clearly distinguished. As it turned out, these forms have genetic differences, but samples 4a and 4b are also not identical in the ITS1-ITS2 section sequences. In this case, it is only possible to estimate which form is closer to *B. cernua* and which one to *B. frondosa* using statistical methods. The plant collected on the territory of Sviblovo Estate in Moscow and based on a set of morphological features defined as *B.* × *decipiens*, in the section ITS1-ITS2, has a very high

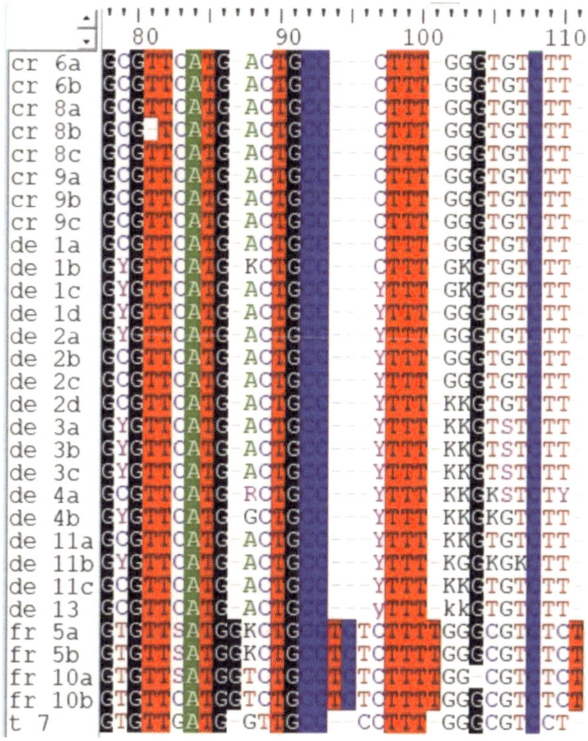

Figure 2.
Fragment of ITS1-ITS2 site of nuclear DNA of various taxa of Bidens *genus. The nucleotides are coded using the International Union of Pure and Applied Chemistry (IUPAC) nomenclature.*

similarity with *B. cernua*. However, in one case this specimen still has heterozygosity in nucleotide substitutions differentiating *B. cernua* and *B. frondosa*, so we cannot say that this specimen is a form of *B. cernua*; most likely, it is the result of introgressive hybridization (**Figure 2**). It is possible that in the case of introgressive hybridization, not only *B.* × *decipiens* × *B. cernua* but also backcrosses are formed (*B.* × *decipiens* × *B. frondosa*). It is interesting that ambiguous readings of a certain nucleotide are also observed for all *B. frondosa* samples in the same position, but they are not related to nucleotide substitutions in other taxa (**Figure 2**). It is not excluded that *B. frondosa* itself is a species of hybrid origin. This is indirectly proved by the high polymorphism of this species in its natural area.

Based on the nucleotide sequences of the ITS1-ITS2 site in the SplitsTree program, the dendrogram is built using the UPGMA method (**Figure 3**). With high probability (with 100% bootstrap support), two clades were separated—sample t_7 (*B. tripartita*). One clade was separated by species, specimens *B. frondosa* (fr_5a, fr_5b, fr_5b, fr_10a, fr_10b), and the other clade included all samples of *B.* × *decipiens* and *B. cernua*, indicating a high similarity.

For trnL-trnF site of chloroplast DNA, samples of *B.* × *decipiens* and *B. cernua* have no differences (this applies to all plants, including those collected in different regions), while *B. frondosa* differs from these taxa by six substitutions of one to two nucleotides and deletion of seven nucleotides (**Figure 4**). *B. tripartita* has another deletion (**Figure 4**), which is absent in other taxa, which once again indirectly confirms its non-participation to the hybrid origin of *B.* × *decipiens*. This means that the aboriginal *B. cernua* is the maternal species and *B. frondosa* is the paternal species.

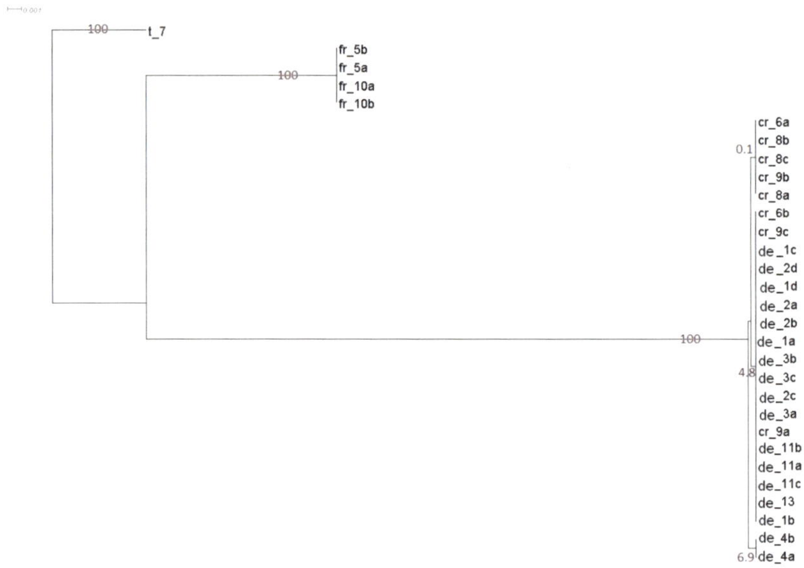

Figure 3.
Dendrogram based on analysis of the ITS region of DNA of various Bidens *taxa with bootstrap support data.*

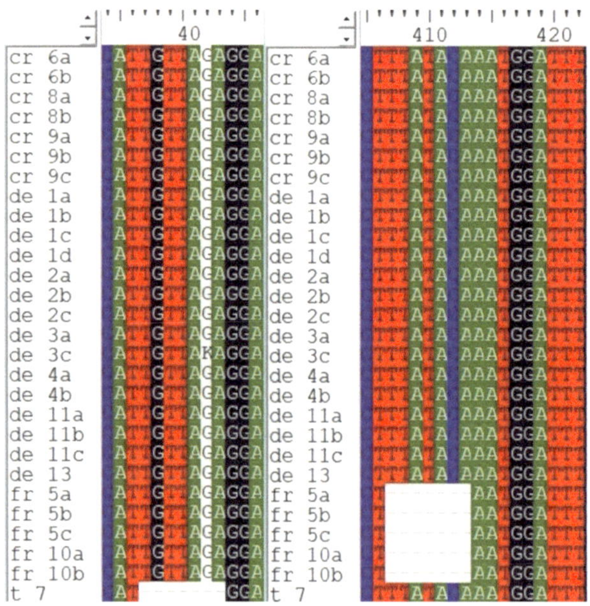

Figure 4.
Fragment of the trnL-trnF intergenic spacer of chloroplast DNA of various taxa of Bidens *genus. The nucleotides are coded using IUPAC nomenclature.*

4. *Solidago* × *niederederi*

We have set ourselves the task to study the features of *Solidago* hybrids in northwest Russia, because in contemporary publications the most numerous references to *S.* × *niederederi* are in the northeastern part of Europe and the Pskov region is the closest region of Russia to it. Previously, on the basis of the analysis of the highly variable noncoding chloroplast region rpl32-trnL by Polish botanists, it was found that hybridization between *S. canadensis* and *S. virgaurea* can occur in both directions and both species can be both mother and father plants [28]. We aimed to determine the situation with respect to parental taxa in the Pskov populations of these species.

The main difference between the hybrid *S.* × *niederederi* and its parents is the structure of shoot systems (mainly the inflorescence structure, **Figure 5**). In *S. canadensis* numerous heads are collected in a compound raceme, and in *S.* × *niederederi* the number of heads is smaller and is collected in a compressed compound raceme, whereas in *S. virgaurea* the number of heads is smaller, and the branches of the compound raceme are so short that the inflorescence is more like a spike.

The size of the heads themselves also varies (**Figure 6**). *S.* × *niederederi* heads have an oval shape and occupy an intermediate position in diameter between parental species, 2201 ± 45 μm (mean ± error average) with a maximum spread of 1762 to 2728 μm, while for *S. virgaurea* and *S. canadensis*, these values are 3132 ± 30 μm (2874–3548 μm) and 1591 ± 22 μm (1428–1939 μm), respectively [29].

With regard to the length of the head, *S.* × *niederederi* plants in Pskov cannot be clearly distinguished from *S. canadensis* due to the high variability of this indicator in *S. canadensis*. However, in terms of average head lengths, the hybrid also occupies an intermediate position between parent species (**Figure 6**). *S. canadensis* and *S.* × *niederederi* shoots are pubescent, while *S. virgaurea* shoots are glabrous, glossy, and sometimes reddish. The leaves of *S.* × *niederederi* in the middle part of the shoot are linear-lanceolate and dentate along the edge, with three distinct veins (as in *S. canadensis*), while in the basal part of the shoot large, ovate, with reticulate veins (as in *S. virgaurea*). To confirm hybrid origin of S. × *niederederi* population in the vicinity of Pskov, nucleotide sequences of nuclear and chloroplast DNA of Pskov individuals (both parent and hybrid species) as well as individuals of parent species from Moscow region were analyzed. The analysis of the ITS1-ITS2 site showed that in all cases of nucleotide substitutions differentiating *S. virgaurea* and *S. canadensis*, *S.* × *niederederi* has ambiguous readings (**Table 2**), indicating heterozygosity, which confirms the hybrid origin of individuals from this population. One sample of *S. canadensis* (c_3c) showed heterozygosity in three cases of nucleotide substitutions out of four, although morphologically this sample did not differ from other individuals of *S. canadensis*, which indicates the presence of introgressive hybridization

Figure 5.
Panicles of Solidago × niederederi *and its parental species.*

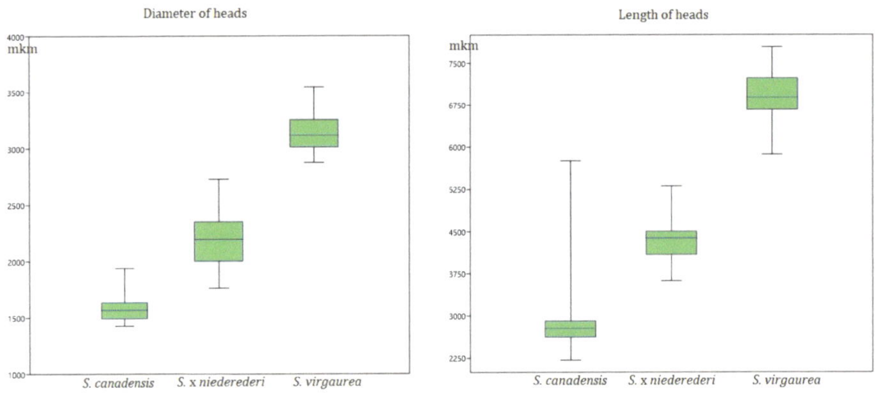

Figure 6.
Parameters of flower heads of Solidago × niederederi *and its parental species: quartiles (first and third), median, maximum, and minimum values are indicated.*

Sample no.	Position in the alignment			
	384	431	508	549
v_1a	T	A	C	A
v_1b	T	A	C	A
v_1c	T	A	C	A
n_2a	Y	M	Y	R
n_2b	Y	M	Y	R
c_3a	C	C	T	G
c_3b	C	C	T	G
c_3c	Y	M	T	R
c_5a	C	C	T	G
v_6a	T	A	C	A
v_6b	T	A	C	A

The nucleotides are coded using IUPAC nomenclature.

Table 2.
ITS1-ITS2 polymorphism for the Solidago × niederederi *hybrid and parental species.*

within the *Solidago* genus. It is likely that this sample is a backcross (result of crossbreeding *S.* × *niederederi* with the parent species *S. canadensis*).

The analysis of the rpl32-trnL high-variable intergenic spacer made it impossible to give an unambiguous answer which species is maternal to *S.* × *niederederi* and which is paternal. In contrast to the results obtained by A. Plizhko and J. Zalevska-Galosh [30], our samples of *S. canadensis* have a higher variability of this section of chloroplast DNA. For example, a sample of s_3a from the Pskov region has a DNA fragment that is absent in other plants of *S. canadensis* not only in the Pskov region but also near Moscow (242–264 nucleotides, **Table 3**). The area occupying positions 271–306 in the alignment of our sequences (292–330 for sample c_3a, **Table 3**) and differentiating parental species in Polish populations [30] may differ not only in *S.* × *niederederi* but also in both parental species. The analysis of another noncoding site of chloroplast DNA, trnL-trnF, also failed to answer this question because all the samples examined were identical in this site. Based on the data obtained, we can only

Sample no.	Position in the alignment				
	191	242–264	271–306 (292–330)	739–741 (746–748, 709–714)	894 (900, 923)
v_1a	A	—	TGTCTAAAAGAATAATTCTTGTATTTCTT	T	C
v_1b	C	—	TGTCTAAAAGAATAATTCTTGTATTTCTTGAATTCT	T	C
v_1c	C	—	TGTCTAAAAGAATAATTCTTGTATTTCTTGAATTCT	—	C
n_2a	C	—	TGTCTAAAAGAATAATTCTTGTATTTCTTGAATTCT	—	C
n_2b	C	—	TGTCTAAAAGAATAATTCTTGTATTTCTTGAATTCT	T	C
n_2c	C	—	TGTCTAAAAGAATAATTCTTGTATTTCTTGAATTCT	T	C
c_3a	A	GAATCTTAATGTTATGTCTAAA	TGTCTAAAAGAATAATTCTTGTATTTCTT	T	A
c_3b	A	—	TGTCTAAAAGAATAATTCTTGTATTTCTTGAATTCT	TTTT	A
c_3c	C	—	—	—	C
v_4	C	—	TGTCTAAAAGAATAATTCTTGTATTTCTTGAATTCT	TT	C
c_5a	C	—	—	T	C
c_5b	A	—	—	TTTT	A

The nucleotides are coded using IUPAC nomenclature.

Table 3.
Polymorphism of the rpl32-trnL region for the Solidago × niederederi hybrid and parental species.

assume that hybridization occurs in both directions in the Pskov population, but there is also a possibility that only one species may be maternal and the other paternal, and it is necessary to search for other, more variable sites of chloroplast DNA.

5. *Erigeron* sect. *Conyza*

Earlier we studied in detail morphological differences between species of the genus *Erigeron* that grow in Eurasia [30], and they are shown in **Table 4**.

Previously a hybrid of *Erigeron canadensis* and *E. sumatrensis*—*Conyza* × *rouyana* Sennen—was described. A typical specimen of this taxon (P04315552), collected by F. Sennen in Catalonia in 1904, is kept in the herbarium of the Museum of Natural History in Paris [P] [7]. Nevertheless, some botanists did not recognize this hybrid and referred *C.* × *royana* to *E. floribundus* [31], which is now treated as synonymous with *E. sumatrensis*. However, it cannot be excluded that morphological differences in several individuals could have been caused not by hybrid processes but by adverse environmental conditions.

The analysis of the **ITS1-ITS2** site of 16 samples of *Erigeron* sect. *Conyza* (supposed hybrids and parent taxa) confirmed our conclusions about the higher polymorphism of *E. sumatrensis* than *E. canadensis*: ITS1-**ITS2** sites of *E. canadensis* samples were identical, while *E. sumatrensis* has substitutions and ambiguous readings (**Table** 5). As for the supposed hybrids (samples 8a, 8b, 10a, 10b, and 22), only in one case an ambiguous reading of the nucleotides coincides with the substitution differentiating *E. canadensis* and *E. sumatrensis*, which indicates that the hybridization has taken place, but since in other cases the substitutions are identical (**Table 5**) and the supposed hybrids have no ambiguous readings, most likely, the reason for their morphological differences is the high polymorphism of *E. sumatrensis* taxon, to which they can be classified.

Features	*E. canadensis*	*E. bonariensis*	*E. sumatrensis*
Number of heads/ generative shoot	500–600	No more 30	less 500
Diameter of heads, mm	4.8 ± 0.1 × 2.4 ± 0.1	6.1 ± 0.1 × 5.2 ± 0.2	6.6 ± 0.1 × 3.2 ± 0.2 At the base, swollen
Structure of shoot systems	The main shoot is barely branched off and ends in a compound raceme occupying the upper third of the escape	The lower lateral deciduous axes of the inflorescence overturn the main shoot axis; the inflorescence covers the upper third of the shoot	The lower lateral deciduous inflorescences are shorter than the main axis of the shoot; the diamond-shaped compound raceme is half the length of the generative shoot
Shape of leaves	Linear-lanceolate with denticle margin	Almost linear with 3–5 denticles	Lanceolate-oval with a serrated margin
Type of pubescence	The leaves are light green, slightly pubescent, the stem is light green, strongly pubescent	The leaves are gray-green, the pubescent leaves and stems are strongly pubescent with long silvery trichomes	The leaves are dark green, softly pubescent, the stems are grayish with abundant soft pubescence

Table 4.
Diagnostic morphological features of Erigeron *species.*

Sample no.	Taxa	Position in the alignment																	
		65, 129, 136, 137, 238, 472/473, 570/571	72, 87, 249, 404	83, 576/577	95	114	130, 567/568, 568/569	211-212	242	412	430	461, 584/585	469	471/472, 530/531, 558/559	499/500	502/503	520/521	535/536	559/560	598-600
3	*Erigeron sumatrensis*	T	C	A	C	C	T	YY	C	C	G	G	—	C	A	C	R	Y	R	TCT
5a		T	C	A	C	C	T	TC	C	C	R	G	—	C	A	C	R	Y	R	TCT
5b		T	C	A	C	C	T	TC	C	Y	R	G	—	C	A	C	R	C	R	TCT
5c		T	C	A	C	C	T	TC	Y	C	R	G	—	C	A	S	R	Y	R	TCT
6		T	C	A	C	S	T	TC	Y	C	R	G	—	C	A	S	R	Y	R	TCT
16		T	C	A	Y	C	T	YY	C	C	G	G	—	C	A	S	R	C	A	TCT
8a	*E. sumatrensis* × *E. canadensis* (?)	T	C	A	Y	C	T	YY	C	C	R	G	—	C	A	S	G	Y	R	TCT
8b		T	C	A	Y	C	T	YY	C	C	R	G	—	C	A	S	R	Y	R	TCT
10a		T	C	A	Y	C	T	YY	C	C	G	G	—	C	A	S	R	Y	R	TCT
10b		T	C	A	C	C	T	TC	C	C	R	G	—	C	A	S	R	Y	R	TCT
22		T	C	A	Y	C	T	YY	C	C	R	G	—	C	A	Y	R	C	R	TCT
13a	*E. canadensis*	C	T	C	C	C	A	CC	C	C	G	G	T	A	T	C	A	C	A	—
13b		C	T	C	C	C	A	CC	C	C	G	A	T	A	T	C	A	C	A	—
18		C	T	C	C	C	A	CC	C	C	G	A	T	A	T	C	A	C	A	—
19	*Erigeron* sp.	C	T	C	C	C	A	CC	C	C	G	A	T	A	T	C	A	C	A	—
20	*E. canadensis*	C	T	C	C	C	A	CC	C	C	G	A	T	A	T	C	A	C	A	—

The nucleotides are coded using IUPAC nomenclature.
Note: "E. sumatrensis × E. canadensis (?)" – putative hybrids.

Table 5.
ITS1-ITS2 polymorphism for different taxa of Erigeron *sect.* Conyza *in the Mediterranean.*

6. Conclusions

Thus, the obtained data on hybrid activity among the Asteraceae family of invasive species are ambiguous.

Hybrid *B.* × *decipiens* has a low polymorphism. *B. cernua* is the most polymorphic taxon, and we can assume the presence of introgressive hybridization of *B.* × *decipiens* with a maternal species. The analysis of ITS1-ITS2 and trnL-trnF sequences showed that *B.* × *decipiens* is of hybrid origin and its maternal form is an aboriginal sequence of *B. cernua* and the paternal one is probably invasive *B. frondosa*. It is possible that *B.* × *decipiens* in its present form has already appeared by introgression, but no morphological differences between supposed hybrids (Belarusian and Kaluga plants) and supposed backcrosses (plants from Moscow and Kaliningrad region) have been revealed. It should be noted that *B. frondosa* itself may be of hybrid origin.

In the northwest of Russia, the populations of three taxa of *Solidago* genus—an invasive species of North American origin of *S. canadensis*, an indigenous species of *S. virgaurea*, and their hybrid *S.* × *niederederi*—grow together in the vicinity of the city of Pskov, which was confirmed by the sequence analysis of the ITS1-ITS2 site. Since both parents, especially *S. canadensis*, are quite polymorphic taxa, it is impossible to answer unambiguously which of the two species is maternal and which is paternal.

In Southern Europe, the hybriogenic activity of representatives of the genus *Erigeron* is close to zero. The low hybriogenic activity can also be explained by differences in the chromosome set: in *E. canadensis* 2n = 18 and in *E. sumatrensis* 2n = 54 [32].

Our data on the rare occurrence of hybrids in comparison with parental species in the Asteraceae family contradict the hypothesis explaining the success of plant growth in the new homeland by strengthening hybridization processes in the secondary distribution range [1, 2], but this situation may change in the coming decades, so the hybriogenic activity of invasive species requires attention of the scientific community.

Acknowledgements

This study was carried out with partial support from the Russian Foundation for Basic Research, grant no. 18-04-00411.

Conflict of interest

The authors declare no conflict of interest.

Author details

Maria A. Galkina* and Yulia K. Vinogradova
N.V. Tsitsin Main Botanical Garden, Russian Academy of Sciences, Moscow, Russia

*Address all correspondence to: mawa.galkina@gmail.com

References

[1] Elton C. The Ecology of Invasions by Animals and Plants. London: Methuen; 1958. p. 181

[2] Ellstrand N, Shierenbeck K. Hybridization as a stimulus for the evolution of invasiveness in plants? Euphytica. 2006;**148**:35-46. DOI: 10.1007/s10681-006-5939-3

[3] Abbott R, James J, Milne R, Gillies A. Plant introduction, hybridization and gene flow. Philosophical Transactions of the Royal Society. 2003;**358**:1123-1132. DOI: 10.1098/rstb.2003.1289

[4] Bleeker W, Schmitz U, Ristow M, et al. Biological Conservation. 2007;**137, 2**:248-253. DOI: 10.1016/j.biocon.2007.02.004

[5] Zalapa J, Brunet J, Guries R. Patterns of hybridization and introgression between invasive *Ulmus pumila* (Ulmaceae) and native *U. rubra*. American Journal of Botany. 2009;**96**: 1116-1128. DOI: 10.3732/ajb.0800334

[6] Albuquerque G, Tauber C, Tauber M. Postmating reproductive isolation between *Chrysopa quadripunctata* and *Chrysopa slossonae*: Mechanisms and geographic variation. Evolution. 1996; **50**:1598-1606. DOI: 10.2307/2410896

[7] Yu V, Maiorov S. Duration of lag-phase as a reflection of microevolution of the plants in secondary range. In: Materialy XIII Moskovskogo soveshchaniya po filogenii rastenii (Proc. XIII Moscow Conf. on the Plant Phylogeny); 2–6 February 2015. Moscow: Maks Press; 2015. pp. 70-74

[8] Williamson M. Biological Invasions. London: Chapman and Hall; 1996. p. 33

[9] Heywood V. Patterns, extents and modes of invasions by terrestrial plants. In: Biological Invasions: A Global Perspective. Chichester: Wiley; 1989. pp. 31-60

[10] Tokhtar VK, Vinogradova YK, Groshenko AS. Microevolution and invasiveness of *Oenothera* L. species (subsect. *Oenothera*, Onagraceae) in Europe. Russian Journal of Biological Invasions. 2011;**2**(4):273-280. DOI: 10.1134/S2075111711040138

[11] Huxel G. Rapid displacement of native species by invasive species: Effects of hybridization. Biological Conservation. 1999;**89**:143-152. DOI: 10.1016/S0006-3207(98)00153-0

[12] Levin D, Francisco-Ortega J, Jansen R. Hybridization and the extinction of rate plant species. Conservation Biology. 1996;**10**:10-16. DOI: 10.1046/j.1523-1739.1996.10010010.x

[13] Rhymer JM, Simberloff D. Extinction by hybridization and introgression. Annual Review of Ecology and Systematics. 1996;**27**: 83-109. DOI: 10.1146/annurev.ecolsys.27.1.83

[14] Yu V, Maiorov S, Khorun L. Chernaya kniga flory Srednei Rossii (The Black Book of Flora of Central Russia). GEOS: Moscow; 2010. p. 502

[15] Rogers S, Bendich A. Extraction of DNA from milligram amounts of fresh, herbarium and mummified plant tissues. Plant Molecular Biology. 1985;**5**: 69-76. DOI: 10.1007/BF00020088

[16] Shaw J, Lickey E, Schilling E, Small R. Comparison of whole chloroplast genome sequences to choose noncoding regions for phylogenetic studies in angiosperms: The tortoise and the hare III. American Journal of Botany. 2007;**94**(3):275-288. DOI: 10.3732/ajb.94.3.275

[17] Strother J, Weedon R. *Bidens* Linnaeus. In: Flora of North America Editorial Committee, editor. Flora of

North America, North of Mexico. Vol. 21. Magnoliophyta: Asteridae, Pt. 8: Asteraceae, P. 3. Oxford: Oxford University Press; 2006. pp. 205-206

[18] Sherff E. The genus *Bidens*. Publications of Field Museum of Natural History, Botanical Series. 1937;**16**:16-74

[19] Sherff E. Further notes on the distribution of *Bidens connata* vars. *pinnata* and *gracilipes*. Rhodora. 1962; **64**(757):23-28

[20] Crowe D, Parker W, et al. Hybridization and agamospermy of Bidens in north-western Ontario. Taxon. 1981;**30**(4):749-760

[21] Andreau J, Vila M. Risk analysis of potential invasive plants of Spain. Journal for Nature Conservation. 2010; **18**(1):34-44. DOI: 10.1016/j.jnc.2009.02.002

[22] Mayorov S, Vinogradova Yu. Formation of secondary distribution range and intraspecific variability of *Bidens connata* [thesis]. Pirenopolis: 12th Reunion on Ecology and Management of Alien Plant Invasions; 2013

[23] Global Plants on JSTOR [Internet]. 2019. Available from: https://plants.jstor.org [Accessed: 03 March 2019]

[24] Galkina M, Vinogradova Y. On the issue of hybridogenic origin of *Bidens* × *decipiens* Warnst. Russian Journal of Biological Invasions. 2019;**10**(4): 315-324. DOI: 10.1134/S2075111719040039

[25] Yu V, Galkina M. On the possibility of hybrid origin of *Bidens connata*. In: Materialy XIII Moskovskogo soveshchaniya po filogenii rastenii (Proc. XIII Moscow Conf. on the Plant Phylogeny); 2-6 February 2015. Moscow: Maks Press; 2015. pp. 64-69

[26] Galkina M, Ryabchenko A. Species of the genus *Bidens* on the banks of the Milyatinskoye reservoir in the Kaluga oblast. Vestn. Tver. Gos. Univ., Ser. Biol., 2013;**32**, **31**:75-83

[27] Galkina M, Vinogradova Yu, Shantser I. Biological and morphological features and microevolution of invasive species of the genus *Bidens* L. Izv. Akad. Nauk, Ser. Biol., 2015;**4**:382-392. DOI: 10.1134/S1062359015040056

[28] Pliszko A, Zalewska-Gałosh J. Molecular evidence for hybridization between invasive *Solidago canadensis* and native *S. virgaurea*. Biological Invasions. 2016;**18**:3103-3108. DOI: 10.1007/s10530-016-1213-3

[29] Galkina M, Vinogradova Y. Invasive taxa of the genus *Solidago* L. in the vicinity of the city of Pskov. Ecosystem Transformation. 2019;**2**(2):12-18. DOI: 10.23859/estr-190207

[30] Vinogradova Y. Sravnitelny analiz biomorfologicheskihk priznakov invazionnykh vidov roda *Conyza* Less. Bulletin of the Main Botanical Garden. 2012;**3**:46-50

[31] Marshall J. *Conyza* – Taxa found in Britain. Watsonia. 1973;**9**:372-373

[32] Krivenko D, Kazanovsky S, Yu V, Verkhozina A, Knyazev M, Murtazaliev R. Chromosome count (Asteraceae). Taxon. 2017;**66**(6): 1491-1492. DOI: 10.12705/666.30

Chapter 7

Changing Climate and Advances on Weeds Utilization as Forage: Provisions, Nutritional Quality and Implications

Muhammad Aamir Iqbal, Sajid Ali, Ayman El Sabagh, Zahoor Ahmad and Muzammil H. Siddiqui

Abstract

Under changing climate, growth and prevalence of many invasive and indigenous weeds are expected to boost up owing to their greater genetic diversity, competitive superiority and better plant architecture. Atmospheric CO_2 enrichment and elevating global temperature are causing weeds invasion to new localities making prevalent weed management strategies ineffective. Weed utilization as forage for ruminants provided that their nutritional profile is available and can be a biologically feasible and economically viable approach compared to existing management system of eliminating them from agro-ecological systems. Different weeds like Bermuda grass (*Cynodon dactylon*), Johnson grass (*Sorghum halepense*), canary grass (*Phalaris minor*), nut sedge (*Cyperus rotundus*), yellow duck (*Rumex crispus*), drooping brome (*Bromus tectorum*), burr clover (*Medicago polymorpha*), button weed (*Diodia scandens*), and purslane (*Portulaca oleracea*) had acceptable nutritional profile with organic matter (89.0–91.3%), protein (7.1–19.5%) and fats (2.1–3.7%). Those were also rich in micro-nutrients (calcium, magnesium and zinc), while anti-nutritional factors (saponins, tannins, phytates and oxalates) were in safer limits for dairy animals. Lack of nutritional profiling and presence of anti-nutritional factors decreased feed intake and led to malnutrition, while higher concentration of tannins caused digestibility depression in small ruminants. There is need to conduct further studies for nutritional profiling of local weed species and development of techniques for reducing their anti-nutritional factors.

Keywords: anti-nutritional factors, global warming, protein content, saponins, tannins

1. Introduction

Climate change has been feared to incur frequent drought spells and floods, while temperature fluctuations and shifting of rainfall patterns are projected to alter growth habits of weeds flora globally. There is an emerging rhetoric that most of invasive and indigenous weeds have the potential and botanical superiority to adjust and acclimatize to atmospheric CO_2 enrichment through the optimization of photosynthesis process leading to significant boost in their biomass production.

In this way, some of the weed species can increase their establishment and dominance in indigenous agro-ecosystems along with invading adjacent as well as far flung terrestrial ecosystems. Thus contrary to notion that weeds are menace and agro-ecological systems must be kept free of indigenous and invasive weeds, their utilization as forage for dairy animals has the potential to become the most feasible and pro-environment strategy [1–3].

Globally, large ruminant's performance is directly influenced by the nutritional value of feed which accounts for over 50% of total expenditures. It deserves mentioning that dairy animals confront forage shortage owing to temperature extremes leading to drastic fall in milk production especially in developing countries. The shortage of forage and rising population of dairy animals has necessitated identifying and evaluating alternate feed resources which are cheap and can also fulfill animal's dietary needs. It has been established that weeds can inflict drastic influence on crops productivity and use of chemical herbicides for keeping them below threshold level, has led to serious concerns pertaining to their residual persistence in crops, soil and environment. Weeds utilization as animal feed holds potential because these are cheap owing to their abundance on field paths and water channels. Weeds harvested from cropped and non-cropped area may constitute an effective and biologically viable approach to keep weeds below the threshold level. Weeds utilization for feeding animals can also reduce herbicides use in agricultural lands which has the potential to curb environmental pollution. Many weeds have been reported to be resistant and better adapt to dynamic environmental conditions, and thus making them less prone to drastic impacts of climate change. In addition, it was reported that animals preferred naturally grown mixtures of weeds over crop residues and roughage during dry season. Furthermore, rapid regeneration favors many weed species for their inclusion as a source of vegetable protein in animal's diet [4–10].

Weeds such as Bermuda grass (*Cynodon dactylon*), Johnson grass (*Sorghum halepense*), canary grass (*Phalaris minor*), nut sedge (*Cyperus rotundus*), cheat-grass or drooping brome (*Bromus tectorum*), burr clover (*Medicago polymorpha*) and pigweed (*Amaranthus viridis*) contained organic matter over 90% indicating that these weeds can fulfill the dry matter requirement of animals. In addition, spotted knapweed (*Centaurea stoebe ssp. micranthos*), a weed of rangelands in Northern America was reported to displace local plant species, degraded wildlife habitats, altered biogeochemistry of soil and triggered soil erosion and thus its control through grazing was found to be biologically and economically viable. Similarly, broom snakeweed and medusa-head (*Taeniatherum caput-medusae*) were effectively controlled through controlled grazing with reasonably good palatability [11–17].

Along with substantial quantity, nutritional quality of weeds is of the utmost importance for dairy animals in order to produce milk on sustainable basis. Field bindweed (*Convolvulus arvensis*) and yellow duck (*Rumex crispus*) were reported to have significantly higher protein content (27 and 22% respectively) [11] which were greater than all cereal forages and most of the legumes, while button weed (*Diodia scandens*) contained 7.7% protein [18]. Although, a number of species belonging to *Commelinaceae* family such as climbing dayflower (*Commelina diffusa* L.), tropical spiderwort/wandering jew (*Commelina benghalensis* L.), Asiatic dayflower (*Commelina communis* L.), African dayflower (*Commelina africana* L.), white mouth dayflower/slender dayflower (*Commelina erecta*) are considered weed but constituted a major chunk of animal feed in Tanzania [19], rural regions of Mauritius [20], USA [21] and in Kenya owing to reasonably good palatability [14, 18, 22, 23].

It is pertinent to mention that anti-nutritional factors (saponins, tannins, oxalates, etc.) of weeds constitute as the most crucial concern as far as animal nutrition is concerned. Tick weed (*Cleomea viscose*) recorded safer limits of anti-nutritional factors such as condensed tannins (0.0491%), saponins (0.23%), phytates (1.2%)

and oxalates (3.3%), but unfortunately in-depth studies are lacking in this context. Furthermore, narrow leaf weeds (nut sedge, wild oat, etc.) were recommended to be a good source of fiber, while broad leaf weeds (pigweed, field bindweed, etc.) provided cheap vegetable protein to dairy animals [1, 11, 13, 20].

To date, very few studies have been done to assess the nutritional status, digestibility and intake of indigenous and exotic (invasive) weeds and their utilization in sustainable ruminant's production systems, but not a single study has so far synthesized and evaluated the literature on weeds utilization as forage. This chapter attempts to synthesize as well as assess the potential of weeds for supplementing traditional feedstuffs (forages, crop residues and concentrates) partially without compromising the productivity of large ruminants in terms of milk production. Weeds mineral constituents and anti-nutritional factors and various implications in weeds utilization as animal feed have also been evaluated.

2. Materials and methods

In order to synthesize published findings pertaining to nutritional quality of weeds, search was performed on Google Scholar (http://scholar.google.com) and PubMed (http://www.PubMed.gov) using the below mentioned search strings:

1. Weeds and forages.
2. Weeds nutritional value.
3. Weeds, animal feed.
4. Anti-nutritional factors in weeds.

The search was time-restricted to 2000–2019, however owing to limitation of published findings; it was later on relaxed to 1990–2019.

The research studies were screened based on following criteria;

1. Reporting at least one weed's biomass production under changing climate.
2. Describing one or more nutritional quality parameters such as protein content of indigenous or invasive weeds.
3. Stating anti-nutritional factors of weeds species.
4. Reporting mineral constituents of at least one or more weeds.

The screening process resulted in 55 studies which fully fit in the objectives and selection criteria.

3. Weeds under changing climate

The rising temperature and carbon dioxide level along with the rapidly altering dynamics of rainfall and evaporation are the most important factors for determining management and utilization of weeds under changing climate. Weeds have been reported to have a greater genetic diversity compared to crops and thus can respond positively to agro-environmental changes. Owing to CO_2 enrichment of atmosphere and rising temperature, some of the weed species can invade new geographical localities while making the existing weed management strategies ineffective. In addition, weeds can have superiority over crop plants by virtue of better plant architecture and incorporating nitrogen and carbon in seeds. Rag weed (*Ambrosia artemissifolia*) developed more number of branches and leaf area along with producing greater number of pollens under increased temperature. Similarly, comparatively higher production of spines by Canada thistle (*Cirsium arvense*) in response to elevated CO_2 level was reported [10, 24–27].

In addition, biomass production of bitter vine/American rope (*mikania mikrantha*), creeping oxeye (*Wedelia trilobata*) and Cairo morning glory (*Ipomea cairica*) was enhanced with increasing CO_2 level [28]. Spurred anoda (*Anoda cristata*) gave the highest green biomass at CO_2 fertilization up to 700 ppm and 32°C temperature, while barnyard grass (*Echinochloa crusgalli*) and Indian goose-grass/wire grass/crowfoot grass (*Eleusine indica*) remained non-responsive to elevated temperature and CO_2 concentration. It was concluded that elevated CO_2 effectively enhanced the photosynthetic process even under water limited conditions indicating higher water use efficiency of weeds under drought stress which led to higher to biomass production. However, weeds response to elevated CO_2 and temperature under well watered conditions continues to remain an unexplored aspect which demands further research to determine the physiological plasticity of different weed species [25, 26].

The temperature elevation as a result of global warming is feared to trigger weeds migration. Cogon grass *(Imperata cylindrica*) prickly acacia (*Acacia nilotica*) and witch weed (*Striga asiatica*) were reported to invade cooler areas of Europe owing to global warming, while some of the invasive weed species such as mesquite (*Prosopis juliflora*) can become more hardy and difficult to control owing to greater portioning of assimilates to roots under elevated temperature particularly under agro-ecological conditions of Indo-Pak subcontinent [27, 29–32].

4. Nutritional quality of weeds

Although, weeds presence in and around the cultivated fields has never been deemed desirable, but these can contribute significantly to the production of quality organic feed for dairy animals. The nutritional profile of weeds determines feasibility and scope for their inclusion in ruminant's feed. The nutritional quality of weeds encompasses digestibility, chemical composition, energy and extent of presence of anti-nutritional factors and such information can assist to determine the allowable proportion of weeds in ruminant's feed [10, 33–35].

Dry matter digestibility has direct relationship with the quality of forage. Different weeds such as Barnyard grass (*Echinochloa crusgalli*) and Jerusalem artichoke (*Helianthus tuberosus*) had significantly higher digestibility compared to many cereal forages. The comparative dry matter digestibility of many weeds and forages crops is presented in **Table 1**. Digestibility was reported to be an important indicator of any forage's quality, while Lamb-squarters (*Chenopodium album*), barn-yard-grass (*Echinochloa crus-galli*), dandelion (*Taraxacum officinale*), Jerusalem artichoke (*Helianthus tuberosus*), yellow foxtail (*Setaria glauca*), perennial sowthistle (*Sonchus arvensis*) and Canada thistle (*C. arvense*) had in-vitro dry matter digestibility equal to alfalfa (*Medicago sativa*). In addition, common ragweed (*Ambrosia artemisiifolia*) and redroot pigweed (*Amaranthus retroflexus*) had even greater in-vitro dry matter digestibility compared to alfalfa [11, 39–41]. There is dire need to determine the digestibility of local weed flora in order to find out their suitability as an alternate animal feed.

Protein (CP) is the most important nutritional quality attribute having direct impact on milk production. Testing of 102 weed species belonging to *Poacea*,

Weeds	DMD (%)	Forage crops	DMD (%)
Barnyard grass (*Echinochloa crusgalli*)		Alfalfa (*Medicago sativa*)	64–75
Canada thistle (*Cirsium arvense*)	68–74	Sorghum (*Sorghum bicolor*)	59–61
Dandelion (*Taraxacum officinale*)	78–84	Maize (*Zea mays*)	63–68
Sowthistle (*Sonchus arvensis*)	76–82	Oat (*Avena sativa*)	60–63
Swamp smartweed/knotweed/tanwed (*Polygonum amphibium*)	54–62	Barley (*Hordeum vulgare*)	59–64
Quackgrass (*Elymus repens*)	58–68	Pearl millet (*Cenchrus americanus*)	58–60
Brome grass (*Bromus tectorum*)	66–76	Cowpea (*Vigna unguiculata*)	68–76
Curly dock (*Rumex crispus*)	50–58	Soybean (*Glycine max*)	70–76
Jerusalem artichoke (*Helianthus tuberosus*)	81–86	Cluster bean (*Cyamopsis tetragonoloba*)	70–79

Table 1.
Dry matter digestibility of some weeds and common forage crops grown under varied agro-climatic conditions [11, 28, 33, 36–38].

Asteraceae, *Fabaceae* and *Euphorbiaceae* families commonly found in central Mexico revealed that only 25 had balanced nutritional profile. Weeds CP content depend on growth stage as matured weeds recorded lesser protein compared to harvestings done at pre-bloom stage. Asthma plant (*Eurphobia hirta*) recorded 16.7% protein content while tick weed or Asian spider-flower (*Cleomea viscose*) with 14.7% followed it, while yellow nutsedge or nut grass (*Cyperus esculentus*) and button weed (*Diodia scandens*) contained 9.8 and 7.7% CP respectively. Bluegrass (*Poa annua*) was found to have over 14% which is higher than maize, sorghum and oat, while common purslane (*Portulaca oleracea*) (8%) was also suggested to be an equally good forage weed as far as CP content is concerned. Another study suggested that weeds including bush sunflower (*Simsia amplexicaulis*), creeping false holly (*Jaltomata procumbens*) and mosquito flower weed (*Lopezia racemosa*) contained CP in the range of 6.5–16.9% and could be used solely or as supplementary feed mixed with maize straw to feed dairy cattle. Mixtures of weeds (*Commelinaceae* + *Amaranthaceae*) recorded crude protein twice than most of the roughages. Another study reported that bush sunflower (*Simsia amplexicaulis*) weed supplemented with maize straw based animal diets resulted in higher protein content successfully met dairy animals dietary needs. Similarly, climbing dayflower (*Commelina diffusa* L.) recorded appreciably higher content of protein (17.7%) which is comparable to commonly used forage crops. In addition, its rumen degradability of protein was recorded over 72% making it forage with balanced nutrition [11, 36, 42–44].

Higher content of fiber increases the bulkiness of feed which results in reduced intake. The lowest crude fiber content was recorded by button weed (18.7%) and nut grass yielded the highest fiber (27%). The minimum lignin content (9.6%) of asthma plant favored its inclusion in animal feed [18]. Similarly, common dandelion (*Taraxacum officinale*) recorded significantly lower crude fiber content (15%), while Bermuda grass (*Cynodon dactylon*) gave the lowest fiber content of just over 6% [11]. In contrast, barnyard grass (*Echinochloa crusgalli*) recorded the highest neutral detergent fiber (NDF) compared to trans-pecos drymary (*Drymaria laxiflora*) [45]. It was reported that climbing dayflower (*Commelina diffusa* L.) recorded 36% and 22% NDF and acid detergent fiber (ADF) respectively and thus compares well to commonly used grasses such as sorghum-Sudan grass and napier grass (*Pennisetum purpureum*) [9]. Similarly, spiderwort (*Tripogandra purpuracens*), a weed of South America recorded reasonably good concentration of carbohydrates which was higher compared to *Tridax coronopifolia* and was recommended to be fed to dairy animals [10, 11, 13, 19, 37, 46].

Digestibility is an important indicator of any forage's quality. Lamb-squarters (*Chenopodium album*), barn-yard-grass (*Echinochloa crus-galli*), dandelion (*Taraxacum officinale*), Jerusalem artichoke (*Helianthus tuberosus*), yellow foxtail (*Setaria glauca*), perennial sowthistle (*Sonchus arvensis*) and Canada thistle (*C. arvense*) had in-vitro dry matter digestibility equal to alfalfa (*Medicago sativa*). In addition, common ragweed (*Ambrosia artemisiifolia*) and redroot pigweed (*Amaranthus retroflexus*) had even greater in-vitro dry matter digestibility compared to alfalfa. Similarly, it was reported that high protein and low fiber contents are indicative of high energy and high productive value feeds. Field bindweed (*Convolvulus arvensis* L.) and common amaranth (*Amaranthus retrofl* exus L.) recorded the highest protein (18.8 and 13.0% respectively) and the lowest fiber (14.7 and 17.6% respectively) which was comparable to alfalfa (*Medicago sativa*) hay having 16.9% protein and 27% fiber. In addition, especial emphasis was paid to palatability of weeds as high nutritional value becomes irrelevant if animals have little likelihood for the weeds species. The hay of different weeds was given to sheep to determine their palatability by using cafeteria of manger technique and biomass consumed in 15 minutes was recorded. Alfalfa had the highest palatability followed by field bindweed (*Convolvulus arvensis* L.)

Weeds	OM (%)	CP (%)	ADF (%)	NDF (%)	F (%)	A (%)
Knapweed (*Centaurea stoebe*) [43]	—	19.5	—	29.5	—	—
Nut sedge (*Cyperus rotundus*) [11]	91.03	16.3	57.8	64.5	—	12.8
Red dead-nettle (*Lamium purpureum*) [43]	—	9.7	25.8	—	2.1	9.8
Field bindweed (*Convolvulus arvensis*) [11]	90.30	27.0	41.0	35.5	—	10.4
Pigweed (*Amaranthus viridis*) [11]	91.00	26.2	57.7	31.0	—	13.2
Johnson grass (*Sorghum halepense*) [46]	—	5.3	30.2	—	1.5	5.5
Field mustard (*Brassica rapa*) [46]	—	9.8	49.5	63.7	—	—
Chicory (*Cichorium intybus*) [46]	—	7.1	35.2	—	3.34	7.5
Bermuda grass (*Cynodon dactylon*) [11]	90.90	13.5	47.0	76.5	—	13.3
Mexican aster (*Cosmos bipinnatus*) [40]	—	10.5	43.5	41.7	—	—
Tick clover (*Desmodium molliculum*) [40]	—	16.2	42.9	41.6	—	—
Yellow foxtail (*Setaria glauca*) [10]	—	20	30.0	—	—	—

Table 2.
Nutritional quality (organic matter, crude protein CP, acid detergent fiber ADF, neutral detergent fiber NDF, fats F, total ash A) of weeds [11, 17, 21, 31, 35, 39, 41, 48, 49].

and common amaranth (*Amaranthus retrofl* exus L.) owing to higher protein and lesser fiber contents. Thus, it was inferred that protein and fiber content of feeds can be used as predictors of palatability and it was also concluded that weeds leaves had 2–3 times higher protein than stems and thus leafy weeds such as field bindweed (*Convolvulus arvensis* L.) recorded higher palatability [11, 45, 47]. The nutritional quality of some weeds has been presented **Table 2**.

5. Mineral constituents of weeds

Minerals in appropriate quantity are essential for dairy animals to be utilized in various metabolic processes, for boosting immunity level against diseases and reproductive health. Asthma plant (*Eurphobia hirta*) was recommended to be included in animal feed for having reasonably higher concentrations of major minerals including calcium (Ca) (13.6%), magnesium (Mg) (3.0%) and potassium (K) (2.5%), along with many trace elements such as iron (Fe) (0.7%), copper (Cu) (0.1%) and manganese (Mn) (0.1%). Common chicory (*Cichorium intybus*) was also suggested as forage weed for having a comparable mineral composition including Ca (6%), Mg (2%), Fe (0.5%) and Cu (0.06%). In addition, pink sorrel (*Oxalis debilis*) was found to be poor on animal nutrition scale for being deficient in Ca (4%), Mg (2.3%), Fe (2.4) and Zinc (Zn) (0.15) compared to other forage weeds [10, 21, 48]. Very limited information has been reported so far regarding mineral constituents of weeds which limit their utilization as a feed source for ruminants. **Table 3** contains mineral constitution of some weeds.

Weeds	Ca	Mg	Zn
Wild oat (*Avena fatua*) [11]	1.8	1.10	0.06
Burr clover (*Medicago polymorpha*) [11]	10.2	2.42	0.14
Morning glory (*Ipomoea purpurea*) [40]	9.0	0.63	2.99
Yellow duck (*Rumex crispus*) [11]	4.7	2.70	0.20
Cheese weed (*Malva parviflora*) [40]	19.3	1.22	4.59
Wood sorrel (*Oxalis decaphyllai*) [40]	5.1	1.43	2.76

Table 3.
Mineral constituents (calcium Ca, magnesium Mg and zinc Zn) of different weeds grown under varied agro-climatic conditions [11, 18, 22, 32, 40].

6. Anti-nutritional contents of weeds

Condensed tannins, saponins, phytate and oxalate are some of the anti-nutritional factors which reduce the nutritional quality and even impart toxicity to animal feeds. It was reported that button weed (0.029–0.052%) had the lowest tannin content, while nutsedge recorded the maximum tannin content. It was suggested that weeds having tannins 2–4% of dry matter did not pose a life threatening situation rather were found to be effective in improving protein flow towards duodenum which led to higher weight gain and reduced the parasitic infections. It was suggested that effective drying of weeds has the potential to significantly reduce the condensed tannins of weeds. Similarly, saponins which are generally produced by defense systems of weeds in response to pathogenic attacks impart a bitter taste and reduce nitrogen digestibility leading to lower palatability of weeds. Tick weed and nutsedge recorded similar saponins (0.22%), while button weed contained higher saponins (0.35%).

Phytates produce phytic acid which acts as a chelator of various macro-minerals (calcium and magnesium) and trace mineral (iron and zinc) leading to a severe deficiency of these minerals. Button weed with 1.18% phytates remained superior to nut-sedge and asthma plants. Oxalate is another important anti-nutritional factor which binds with calcium to form calcium oxalate leading to calcium unavailability. Asthma plant recorded the lowest oxalate concentration (2.36%) while button weed had the maximum oxalate concentration (2.92).

It was suggested that weeds such as Arizona sunflower (*Tithonia tubiformis*), wood sorrel (*Oxalis divergens*) and bush sunflower (*Simsia amplexicaulis*) commonly found in America, Mexico, Argentina and Chile contained tannins in safer limits and might be utilized to feed dairy animals. In contrast, a fatty acid called malvalic acid was isolated from cheese weed (*Malva parviflora* L.) which caused deaths of dairy animals. Similarly, different phenolic compounds were reported to be the major reason behind low palatability of many weed species. In addition, the presence of phytochemicals and free oxygen metabolites in weeds contributed to mastitis and ultimately led to udder edema along with deteriorating the reproductive performance of cattle. Similarly, spotted knapweed (*Centaurea stoebe ssp. micranthos*) in North American rangelands contained an allelo-chemical named cnicin (a sesquiterpene lactone compound) which reduced its palatability by imparting bitter taste and deterred grazing. In contrast, knapweed was readily consumed by small ruminants' preferably at rosette and bolting stages compared to flowering and seed-set phonological stages. This preference was associated with higher protein and lower fiber content at rosette stage in comparison to flowering or seed-set stages without any link between cnicin content and knapweed palatability [28, 49–51].

7. Nutritional comparison of weeds and forage crops

To the best of our knowledge, no comprehensive studies have been reported pertaining to qualitative analyses of different native and exotic weeds with forage crops. Some of the weeds such as Canada thistle, spotted knapweed, white-top, Russian knapweed and pigweeds contain protein in the range of 15–22% while typical grasses has only 2–11% protein, thus have the potential to become cost-free source of plant protein. In addition, higher leaf-stem ratio in weeds impart them superiority over grasses in terms of higher digestibility. Moreover, weeds provide nutrients rumen microbes which enable dairy animals to digest lower quality forages and thereby reducing overall feed cost. Weeds are always available even during periods of drought in arid areas and their utilization can help to obtain sustainable supplies of milk throughout the years with minimum cost. Dairy animals being fed on protein rich weeds tend to gain weight more rapidly and that too with no additional cost.

8. Limitations and implications

The studies regarding nutritional quality, presence of anti-nutritional or toxic substances and palatability of most of the native weed species is lacking. However, one of the most important limitations in utilizing tropical weeds for dairy animals is the presence of anti-nutritional factors such as tannins which are harmful and toxic to ruminants [13, 38, 51]. Animals being fed on weeds having high concentration of tannins witnessed digestibility depression. Weeds having exceptionally higher lignin content caused a sharp decline in feed intake leading to serious malnutrition in dairy animals. Similarly, significantly less palatability of weeds reduce their intake, which led to animal's weight loss along with sharp decline in milk production [52–54].

Moreover, some of the weeds species have thorns and spines due to which animal gets its mouths injured along with irritating of eyes which leads to pinkeye. In addition, there could be some weeds which can impart unpleasant odor and taste to milk and meat. Lastly, although weeds offer cost-free source of animal feed but centuries-long war against weeds has made it difficult to change the mind of ranchers and dairy farmers to utilize this precious source of plant protein which needs to be changed.

9. Conclusions

The exceptional resistance to drought, higher biomass production under unfavorable pedo-climatic conditions, rapid regeneration capacity, and acceptable nutritional quality at all phonological stages suggests that there are opportunities to utilize weeds as forage for all types of ruminants. Weeds availability throughout the year warrants their potential to fulfill essential dietary needs of animals and favors their inclusion as supplementary forage especially during extreme weather conditions. However, controlled field investigations for determining the appropriate growth timings, nutritional quality, anti-nutritional factors and biomass production potential of different native weeds must be done in rangelands while maintaining a balanced and healthy rangeland ecosystem.

Author details

Muhammad Aamir Iqbal[1*], Sajid Ali[2], Ayman El Sabagh[3,4], Zahoor Ahmad[5] and Muzammil H. Siddiqui[1*]

1 Department of Agronomy, Faculty of Agriculture, University of Poonch Rawalakot (AJK), Pakistan

2 Institute of Agricultural Sciences, University of the Punjab, Lahore, Pakistan

3 Department of Field Crops, Faculty of Agriculture, Siirt University, Turkey

4 Department of Agronomy, Faculty of Agriculture, Kafrelsheikh University, Egypt

5 Department of Field Crops, Faculty of Agriculture, Cukurova University, Turkey

*Address all correspondence to: muhammadaamir@upr.edu.pk and muzammilagro@upr.edu.pk

References

[1] Ziska LH, Faulkner SS, Lydon J. Changes in biomass and root:shoot ratio of field grown Canada thistle (*Cirsium arvense*), a noxious, invasive weed, with elevated CO_2. Weed Science. 2004;**47**:608-615

[2] Kalita P, Mukhopadhyay PK, Mukherjee AK. Evaluation of the nutritional quality of four unexplored aquatic weeds from Northeast India for the formulation of cost-effective fish feeds. Food Chemistry. 2007;**103**:204-209

[3] Iqbal MA, Muzammil HS, Sher A, Zahoor A, Qaiser M, Rana DK. Forage productivity of cowpea [*Vigna unguiculata* (L.) Walp] cultivars improves by optimization of spatial arrangements. Revista Mexicana De Ciencias Pecuarias. 2018;**9**:203-219

[4] Iqbal MA, Asif I, Muzammil HS, Zahid M. Bio-agronomic evaluation of forage sorghum-legumes binary crops on Haplic Yermosol soil of Pakistan. Pakistan Journal of Botany. 2018;**50**:1991-1997

[5] Iqbal MA, Asif I, Zahid M, Zahoor A, Essa A, Muzammil HS, et al. Revamping soil quality and correlation studies for yield and yield attributes in sorghum-legumes intercropping systems. Bioscience Journal. 2018;**34**:1165-1176

[6] Iqbal MA, Asif I, Rana NA. Spatio-temporal reconciliation to lessen losses in yield and quality of forage soybean (*Glycine max* L.) in soybean-sorghum intercropping systems. Bragantia. 2018;**77**:283-291

[7] Iqbal A, Muhammad AI, Imtiaz H, Muzammil HS, Muhammad N, Jamshed A, et al. Seed blending of oat (*Avena sativa* L.) and canola (*Brassica napus* L.) under variable seed proportions enhanced forage productivity and nutritional quality. Pakistan Journal of Botany. 2018;**50**:1985-1990

[8] Iqbal MA, Abdul H, Imtiaz H, Muzammil HS, Tanveer A, Abdul K, et al. Competitive indices in cereal and legume mixtures in a South Asian environment. Agronomy Journal. 2019;**111**:242-249. DOI: 10.2134/agronj2017.11.0658

[9] Lanyasunya TP, Wang HR, Abdulrazak SA, Mukisira EA, Zhang J. The potential of the weed, *Commelina diffusa* L., as a fodder crop for ruminants. South African Journal of Animal Science. 2006;**36**:27-31

[10] Ziska LH. Evaluation of yield loss in field sorghum from a C3 and C4 weed with increasing CO_2. Weed Science. 2003;**51**:914-918

[11] Khan R, Khan MA, Sultan S, Marwat KB, Khan I, Hassan G, et al. Nutritional quality of sixteen terrestrial weeds for the formulation of cost-effective animal feed. Journal of Animal and Plant Sciences. 2013;**23**:75-79

[12] Thrift BD, Mosley JC, Brewer TK, Roeder BL, Olson BE, Kott RW. Prescribed sheep grazing to suppress spotted knapweed on foothill rangeland. Rangeland Ecology and Management. 2008;**61**:18-25

[13] LeJeune KD, Seastedt TR. Centaurea species: The forb that won the west. Conservation Biology. 2001;**15**:1568-1574

[14] Olson BE, Lacey JR. Sheep: A method for controlling rangeland weeds. Journal of Sheep and Goat Research. 1994;**10**:105-112

[15] Wright AL, Kelsey RG. Effects of spotted knapweed on cervid winter–spring range in Idaho. Journal of Range Management. 1997;**50**:487-496

[16] Olson BE, Kelsey RG. Effect of *Centaurea maculosa* on sheep rumen microbial activity and mass in vitro. Journal of Chemistry and Ecology. 1997;**23**:1131-1144

[17] Ralphs MH, Banks JE. Cattle grazing as a biological control for broom snakeweed: Vegetation response. Rangeland Ecology and Management. 2009;**62**:38-43

[18] Akinfemi A, Mako AA. Assessment of nutritive value of four dominant weed species in north central Nigeria. Livestock Research for Rural Development. 2012;**24**. Available from: http://www.lrrd.org/lrrd24/11/akin24208.htm

[19] Ingratubun GF, Owen E, Massawe NF, Mtenga LA, Mtengeti EG. Effect of upgrading small East African goats on feed resource utilisation in the Uluguru mountains in Tanzania; a farmers' perspective. Livestock Research for Rural Development. 2000;**12**:13-19

[20] Boodoo AA, Ramjee R, Hulman B, Dolberg F, Rowe JB. Evaluation of the basal forage diet of village cows. Livestock Research for Rural Development. 1990;**1990**:2. Available from: http://www.lrrd.org/lrrd24/11/akin24208.htm

[21] Mullahey JJ. Tropical soda apple, wetland nightshade, and Turkey berry. In: Van Driesche R et al., editors. Biological Control of Invasive Plants in the Eastern United States. Florida, USA: USDA Forest Service Publication; 2002. pp. 413-423

[22] Walker JW, Kronberg SL, Al-Rowaily SL, West NE. Comparison of sheep and goat preference for leafy spurge. Journal of Range Management. 1994;**47**:429-434

[23] Aggarwal GC, Pimentel D, Mario G. Weed benefits and costs in rice and wheat production in India. Agricultura, Ecosystems & Environment. 1992;**39**:235-244

[24] Hyvonen T, Glemnitz M, Radics L, Hoffmann J. Impact of climate and land use type on the distribution of finish casual arable weeds in Europe. Weed Research. 2010;**51**:201-208

[25] Kriticos DJ, Sutherst RW, Brown JR, Adkins SW, Maywald GF. Climate change and the potential distribution of an invasive alien plant: *Acacia nilotica* spp. *indica* in Australia. Journal of Applied Ecology. 2003;**40**:111-124

[26] Patterson DT. Weeds in a changing climate. Weed Science. 1995;**43**:685-701

[27] Malarkodi N, Manikandan N, Ramaraj AP. Impact of climate change on weeds and weed management—A review. Journal of Innovative Agriculture. 2017;**4**:1-6

[28] Song L, Wu J, Furong LC, Peng S, Chen B. Different responses of invasive and native species to elevated CO_2 concentration. Acta Ecologica. 2009;**35**:128-135

[29] Chaudhury UN, Kirkham MB, Kanemasu ET. Carbon dioxide and water level effects on yield and water use of winter wheat. Agronomy Journal. 1990;**82**:637-641

[30] Ziska LH. Influence of rising atmospheric CO_2 since 1900 on early growth and photosynthetic response of a noxious invasive weed, Canada thistle (*Cirsium arvense*). Functional Plant Biology. 2002;**29**:1387-1392

[31] Ziska LH, George K. Rising carbon dioxide and invasive, noxious plants: Potential threats and consequences. World Resource Review. 2004;**16**:427-447

[32] Tang J, Xu L, Chen X, Hu S. Interaction between C4 barnyard grass

and C3 upland rice under elevated CO_2: Impact of mycorrhizae. Acta Ecologica. 2009;**35**:227-235

[33] Castelan O, Estrada J, Carretero L, Vieyra A, Martínez N, Cárdenas S, et al. Degradation characteristics of maize weeds used as forage in smallholder maize-livestock production systems of central México, in different growing periods. Tropical Subtropical Agroecosystems. 2003;**3**:115-119

[34] De-Bruijn SL, Bork EW. Biological control of Canada thistle in temperate pastures using high density rotational cattle grazing. Biological Control. 2006;**36**:305-315

[35] Di-Tomaso JM, Kyser GB, Goerge MR, Doran MP, Laca EA. Control of Medusahead (*Taeniatherum caput*-medusae) using timely sheep grazing. Invasive Plant Science Management. 2008;**1**:241-247

[36] Holst PJ, Allan CJ, Campbell MH, Gilmour AR. Grazing of pasture weeds by goats and sheep, pt. 1: Nodding thistle (*Carduus nutans*). Australian Journal of Experimental Agriculture. 2004;**44**:547-551

[37] Lamers J, Burkert A, Makkar HPS, Von Oppen M, Becker K. Biomass production, feed and economic value of fodder weeds as by-products of millet cropping in a Sahelian farming system. Experimental Agriculture. 1996;**32**(3):317-326

[38] Ziska LH, Caulfield FA. Rising CO_2 and pollen production of common ragweed (*Ambrosia artemisiifolia*), a known allergy inducing species: Implications for public health. Australian Journal of Plant Physiology. 2000;**27**:893-898

[39] Di-Tomaso JM. Invasive weeds in rangelands: Species, impacts, and management. Weed Science. 2000;**48**:255-265

[40] Dora G, Mendoza S, Serrano V, Bah M, Pelz R, Balderas R, et al. Proximate composition, mineral content, and antioxidant properties of 14 Mexican weeds used as fodder. Weed Biology and Management. 2008;**8**:291-296

[41] Frost RA, Launchbaugh KL. Prescription grazing for rangeland weed management. Rangelands. 2003;**25**:43-47

[42] Frost RA, Wilson LM, Launchbaugh KL, Hovde EM. Seasonal change in forage value of rangeland weeds in Northern Idaho. Invasive Plant Science and Management. 2008;**1**:343-351

[43] Ganguli AC, Michael BH, Karen LL. Seasonal change in nutrient composition of spotted knapweed and preference by sheep. Small Ruminant Research. 2010;**89**:47-50

[44] Gutman M, Perevolotsky A, Sternberg M. Grazing effects on a perennial legume, *Bituminaria bituminosa* (L.) Stirton, in a mediterranean rangeland. Cahiers Options Mediterranean. 2000;**45**:299-303

[45] Martinez-Loperena R, Castelan-Ortega OA, Gonzalez-Ronquillo M, Estrada-Flores JG. Nutritive value, in vitro fermentation and secondary metabolites of weeds and maize straw used for feeding dairy cattle. Tropical and Subtropical Agroecosystems. 2011;**14**:525-536

[46] Kirilov A, Georgieva N, Stoycheva I. Determination of composition and palatability of certain weeds. International Journal of Agricultural Science and Food Technology. 2016;**2**:41-43

[47] Miller JK, Brzezinska-Slebodzinska E. Oxidative stress, antioxidants, and animal

function. Journal of Dairy Science. 1993;**76**:2812-2823

[48] Olson BE, Wallander RT, Lacey JR. Effects of sheep grazing on spotted knapweed-infested Idaho fescue community. Journal of Range Management. 1997;**50**:386-390

[49] Iqbal MA, Asif I, Nadeem A, Haroon ZK, Rana NA. A study on feed stuffs role in enhancing the productivity of milch animals in Pakistan-existing scenario and future prospect. Global Veterinária. 2015;**14**:23-33

[50] Shale TL, Stirk WA, Van SJ. Variation in antibacterial and anti-inflammatory activity of different growth forms of *Malva parviflora* and evidence for synergism of the anti-inflammatory compounds. Journal of Ethnopharmacology. 2005;**96**:325-330

[51] Tremmel DC, Patterson DT. Responses of soybean and five weeds to CO_2 enrichment under two temperature regimes. Canadian Journal of Plant Science. 1993;**73**:1249-1260

[52] Iqbal MA, Abdul H, Tanveer A, Imtiaz H, Sajid A, Ansar A, et al. Forage sorghum-legumes intercropping: Effect on growth, yields, nutritional quality and economic returns. Bragantia. 2019;**78**:82-95

[53] Iqbal MA, Abdul H, Muzammil HS, Imtiaz H, Tanveer A, Saira I, et al. A meta-analysis of the impact of foliar feeding of micronutrients on productivity and revenue generation of forage crops. Planta Daninha. 2019;**37**:e019189237

[54] Akdeniz H, Mehmet AB, Ibrahim H, Mohammad SI, Akbar H, Mabrouk S, et al. Yield and nutritive value of Italian ryegrass (*Lolium italicum* L.) is influenced by different levels of nitrogenous fertilizer. Fresenius Environmental Bulletin. 2019;**28**:8986-8992

Chapter 8

Valorization of Prickly Pear [*Opuntia ficus-indica* (L.) Mill]: Nutritional Composition, Functional Properties and Economic Aspects

Francisca Hernández García, Lucia Andreu Coll, Marina Cano-Lamadrid, David López Lluch, Ángel A. Carbonell Barrachina and Pilar Legua Murcia

Abstract

Opuntia ficus-indica (L.) Mill, usually named prickly pear or nopal cactus, is the Cactaceae plant with the greatest economic relevance in the world. It is a tropical or subtropical plant, native to tropical and subtropical America, which can grow in arid and semiarid climates. Prickly pear is mainly known by its fruits, popularly named "tunas" or "figs," but their cladodes are also consumed, principally in Mexico, which is the country with the largest cultivated area and the largest producer. There is ample evidence of the health benefits of prickly pear: it shows high antioxidant activity, it is a source of nutrients and vitamins and it presents medicinal uses, among others. Furthermore, prickly pear presents other uses, including cosmetics, biofuel production, animal nutrition and soil phytoremediation.

Keywords: cactus pear, fruit, antioxidant activity, health benefits, peel, pulp

1. Introduction

Opuntia ficus-indica L. Mill, usually known as prickly pear, cactus pear or nopal, is a tropical or subtropical plant that belongs to the Cactaceae family, originally from arid and semiarid regions of America. This plant can grow in arid and semiarid climates, being the Cactaceae plant with the greatest economic relevance in the world [1]. It produces an edible and highly flavored fruit, known as "cactus pear," which is a berry with numerous seeds and thick peel, enclosing a delicately flavored pulp [2]. Their cladodes are also consumed, mainly in Mexico, which is the country with the largest area under cultivation and the largest producer [3, 4] but it is also cultivated in the United States, Spain, Italy, South Africa and Argentina, among other countries [5, 6]. Prickly pear fruit is commonly consumed in fresh, but it can also be consumed as juices, jam, syrups and other processed products. They are widely employed in Latin America. The current demand of prickly pear in Spain is increasing [4, 6].

There are ample evidences of the health benefits of consumption of prickly pear due to its source of nutrients and vitamins [4, 7, 8] and antioxidant properties due to its content of bioactive compounds [2, 9, 10]. Additionally, prickly pear presents medicinal uses: it is used in treatment of hyperglycemia and high levels of cholesterol [7, 11, 12] and its consumption is linked with lower incidence of coronary diseases and some types of cancer [8, 13], among others.

This chapter is focused on the nutritional composition, bioactive compounds and economic aspects of prickly pear fruits through a compilation and synthesis of the available studies. With this, the authors intend to contribute to the knowledge of *O. ficus-indica* and also to promote new scientific research and industrial use of this crop.

2. Nutritional composition

Table 1 shows nutritional composition of prickly pear pulp and peel. Prickly pear fruit pulp has high content of protein, lipids and moisture but low content of total fiber and ash comparing to the peel.

About sugar profile, glucose and fructose are the predominant ones in both peel and pulp. On average, fruit pulp shows high content of glucose (123 g L^{-1}) and fructose (71.7 g L^{-1}) than peel (91.0 and 52.0 g L^{-1}, respectively) [9].

Prickly pear fruit also stands out for its mineral contents. Potassium is the major macronutrient in pulp (199–410.7 mg 100 g^{-1} dw), followed by calcium (12.4–49.1 mg 100 g^{-1} dw) and magnesium (18 mg 100 g^{-1} dw). Fruit peel presents magnesium (18.6–987 mg 100 g^{-1} dw), calcium (49.04–951) and potassium (320–549 mg 100 g^{-1} dw) as the major macronutrients [14, 15]. Fruit pulp shows lower level of sodium (0.70–1.09 mg 100 g^{-1} dw) than peel (1.8–951 mg 100 g^{-1} dw) [14–16]. Iron, manganese and copper are the major microelements in fruit peel and pulp [14, 15]. The mineral pattern depends on the fruit origin and crop factors [15].

Constituents	Unit	Pulp	Peel	References
Moisture	%	90.66	88.92	[15]
Titratable acidity	g citric acid L^{-1}	0.23–1.60	0.61–3.40	[9]
Total soluble solids	° Brix	10.7–15.7	8.03–15.4	[9]
pH	—	5.41–6.01	4.83–5.59	[9]
Energy	kcal 100 g^{-1} dw	361	169	[17]
Protein	% dw	1.62	1.53	[15]
Lipids	% dw	0.56	0.32	[15]
Total fibers	% dw	4.65	5.83	[15]
Ash	% dw	2.60	3.40	[15]
Fructose	g L^{-1}	57.8–88.0	27–81.8	[9]
Glucose	g L^{-1}	103–144	57–128	[18]

Table 1.
Nutritional composition of prickly pear fruit pulp and peel.

3. Bioactive compounds

Table 2 shows the main bioactive compounds present in prickly pear fruit peel and pulp. These are betalains (betanin and indicaxanthin), flavonoids, phenolics, vitamin C and carotenoids.

Compounds	Unit	Pulp	Peel	References
Vitamin C	mg 100 g^{-1} fw	28–79.2	59.8	[2, 14, 19, 20]
Total flavonoids	mg rutin equivalents g^{-1} fw	0.2–0.7	1.4–2.8	[21, 22]
Total phenolic content	mg rutin equivalents g^{-1} fw	2–2.5	5.4–6.2	[21]
Carotenoids	μg g^{-1} fw	2.56–3.79	12.58–16.93	[2, 6]
Indicaxanthin	mg 100 g^{-1} fw	2.61–39.6	—	[19, 20]
Betanin	mg 100 g^{-1} fw	0.10–1.04	—	[20]

fw, fresh weight.

Table 2.
Principal bioactive compounds in prickly pear fruit pulp and peel.

Betalains are water-soluble pigments (containing nitrogen) that are responsible for the red or yellow color of fruits, flowers, roots and leaves of plants belonging to the order of *Caryophyllales*, in which Cactaceae plants are included. [19]. Prickly pear fruits are characterized by various colors due to the combination of two betalain pigments, the purple-red betanin and the yellow-orange indicaxanthin [20]. These compounds make prickly pear fruits a good source of bioactive compounds with antioxidant properties, which may have beneficial effects on the consumer's health [19].

Flavonoids are a group of secondary metabolites of plants implicated in fruit and flower coloration, photosensitization and energy transfer, among others. Flavonoids present high antioxidant activity that helps to neutralize damaging free radicals and to prevent oxidative stress in the human body [21, 22]. Prickly pear fruits contain more flavonoids in the peel than in the pulp and there are fewer flavonoids than phenolic compounds (**Table 2**) [21].

Vitamin C is an essential nutrient for humans that provides a high antioxidant activity and prevents against oxidative stress in humans [14, 20, 21]. The content of this vitamin depends on the cultivar among other factors, being higher in red cultivars, which show higher concentration of vitamin C than some common fruits such as apple, peach and grapes [2].

Carotenoids are organic pigments that belong to isoprenoid group and are widely distributed among fruits. They are responsible for most yellow, orange and red colors in vegetables. These pigments contribute to the appearance and attractiveness of a fruit. They can also perform as antioxidants [2, 6]. Concentration of carotenoids in prickly pear fruits is slightly lower than that reported for other fruits but it confirms the observation that yellow-colored fruits present higher concentrations than colored fruits [2].

4. (Poly)phenols and phenolic profile

Polyphenols are an important group of natural compounds, founded in plants and characterized by the presence of more than one phenol group in their structure. These molecules are considered to be of high scientific and therapeutic interest, because they help to prevent degenerative diseases, cardiovascular diseases and cancers, among others, due to their antioxidant activity [21, 23].

In general, the peel of prickly pear fruits is richer than pulp in total phenolic content [21, 24, 25] (**Table 2**). The profile of individual (poly)phenolic compounds depends on the cultivar [18]. Generally, predominant compounds in prickly pear fruit

Compound	Pulp [18]	Peel [18]	Pulp [5]	Peel [5]	Pulp and peel [26]
Protocatechuic acid-hexoside	x	x			
Piscidic acid			x	x	x
Caffeic acid 4-O-glucuronide					x
4-Hydroxybenzoic acid derivative			x	x	
p-Coumaric acid 4-O-glucoside					x
Myricetin-hexoside	x	x			
Ferulic acid derivative	x	x			x
Ferulic acid-hexoside	x	x			
Guaiacyl(t8-O-4)guaiacyl-hexoside	x	x			
Sinapic acid-hexoside	x	x			
Syrinigyl(t8-O-4)guaiacyl	x	x			
Quercetin-hexoside-pentoside	x				
Quercetin-rhamnose-hexoside-rhamnose		x			
Rutin-pentoside		x			
Syrinigyl(t8-O-4)guaiacyl	x				
Kaempferol-di-rhamnose-hexoside		x			
Kaempferol-glucosyl-rhamnoside			x	x	x
Kaempferol 3-O-(2”rhamnosyl-galactoside)7-O rhamnoside					x
Taxifolin					x
Isorhamnetin-rhamnose-rutinoside	x	x			
Isorhamnetin glucosyl-rhamnosyl-rhamnoside			x	x	
Isorhamnetin glucosyl-pentoside			x	x	
Isorhamnetin glucosyl-rhamnoside			x	x	x
Quercetin-hexoside-pentoside	x	x			
Isorhamnetin derivative	x	x			
Dihydrosinapic acid hexoside	x	x			
Quercetin-3-O-rutinoside (rutin)		x	x	x	
Secoisolariciresinol-hexoside	x	x			
Quercetin-hexoside	x	x			
Kaempferol-rutinoside		x			
Syringaresinol	x	x			
Naringenin-hexoside	x	x			
Isorhamnetin-rutinoside	x	x			
Isorhamnetin-3-O-glucoside					x
Isorhamnetin diglucoside					x
Isorhamnetin-C-hexoside		x			
Eucomic acid					x

Compound	Pulp [18]	Peel [18]	Pulp [5]	Peel [5]	Pulp and peel [26]
Naringin	x	x			
Guaiacyl(8-O-4) syrinigyl (8-8) guaiacyl-hexoside	x	x			
Feruloyl derivative	x				
Trihydroxy-methoxy-flavonol	x	x			

Table 3.
Phenolic compounds found in prickly pear fruit peel and pulp in the most recent studies.

pulp and peel are ferulic acid and derivatives, isorhamnetin and derivatives, sinapic acid and derivatives, and quercetin and derivatives [5, 18, 24]. Other compounds found in these botanical parts are kaempferol, myricetin, luteolin, catechin, naringin and syringaresinol, among others [5, 18, 24].

The presence of the phenolic compounds in prickly pear fruit peel and pulp, due to its antioxidant activity, makes this fruit an important product that can contribute to prevent human degenerative diseases such as cancer, diabetes, hypercholesterolemia, arteriosclerosis or cardiovascular and gastric diseases [21, 25]. **Table 3** shows some compounds found in the most recent studies [5, 18, 26] about phenolic profile of prickly pear fruit peel and pulp.

5. Sugars and organic acid composition

Citric and malic acids are the major organic acids present in prickly pear fruit pulp and peel. Other organic acids, such as oxalic, tartaric, quinic, shikimic and fumaric acids, are present in traces. Citric acid ranges from 1.60 to 3.20 g L^{-1} in fruit peel and shows values from 0.30 to 1.61 in pulp g L^{-1} [9]. Malic acid shows concentrations between 1.04 and 2.20 g L^{-1} in peel and 1.20 and 2.10 g L^{-1} in pulp. However, cladodes show higher values of these acids (71.8 g L^{-1} of malic acid and 37.7 g L^{-1} of citric acid) and also contain succinic acid (43 g L^{-1}) [9]. This is due to the CAM metabolism of *O. ficus-indica*, especially in the cladodes. Organic acids are accumulated in the vacuole during night and suffer a reciprocal reserve carbohydrates accumulation during the daytime phase [27].

Organic acids in fruits are in lower concentration in comparison with cladodes; however, fruits, especially pulp, are characterized by high sugar content. Some authors [9] studied the concentration of glucose and fructose in fruits and their results show that glucose predominates over fructose in both fruit peel and pulp (123 g L^{-1} of glucose and 91 g L^{-1} of fructose in pulp versus 91 g L^{-1} of glucose and 52 g L^{-1} of fructose in fruit peel). However, other studies [28] show that glucose, fructose and sucrose concentration is higher in fruit peel than in pulp. These results indicate that concentration of sugars may depend on the cultivars.

Sugar concentration in prickly pear fruit makes it a good source of energy and a natural source of sweetness for food preparations. Besides, fructose contributes to the typical sweet taste of this fruit, due to its high wetness compared with glucose and sucrose [29].

6. Volatile compounds

Volatile compounds influence the sensory quality of fruits. Their aromas are formed from a complex group of chemical substances such as aldehydes, alcohols, ketones, terpenes and esters, among others. These compounds usually show a low concentration in fruits and their variability depends on cultivar, climatological conditions, maturity and storage conditions, among other factors [30]. In prickly pear fruit pulp, the content of these compounds varies from 3.33 mg 100 g^{-1} to 14.86 mg 100 g^{-1} [31].

Even though prickly pears have no strong aroma, up to 61 compounds have been identified [32]. In a recent research [31], the studied cultivars showed aldehydes and terpenes as the most numerous compounds. Both chemical groups and alcohols were the most abundant compounds. However, other studies reported alcohols [32–34] and esters [35] as the most numerous and abundant compounds. Some predominant compounds are D-limonene (citrus notes), 2,6-nonadienal (vegetable notes), nonanol (green, melon and fatty attributes), 2-hexenal (almond, apple green, sweet and vegetable notes), and 1-hexanol (green and sweet notes), among others [31–33].

Although prickly pear fruits are highly valued for their health-promoting benefits, sensory analysis is needed to complete the knowledge of aroma of this fruit and the effect of the cultivar [31].

7. Fatty acids

The consumption of monounsaturated and polyunsaturated fatty acids (MUFAs and PUFAs, respectively) has been stated to provide health benefits. It also contributes to the improvement of various health conditions regarding obesity, cardiovascular diseases, diabetes mellitus and even some types of cancer [13, 36].

Prickly pear fruit pulp and peel showed important percentages of MUFAs and PUFAs. In fruit pulp, MUFAs ranged from 16.9 to 40.2% (as % of total of fatty acid profile) and PUFAs ranged from 35.2 to 53.9%. Fruit peel showed slightly lower values of MUFAs (6.90–31%) but higher ones in PUFAs (37.0–63.2%) [37]. Furthermore, prickly pear seed oil showed high percentages of PUFAs, recorded at levels between 57.90 and 63.29%, and MUFAs, ranged from 19.81 to 23.30% [38].

The most abundant compounds in fruit pulp, peel and seed oil were linoleic (C18:2), oleic (C18:1) and palmitic (C16:0) acids [15, 37]. Prickly pear fruit peel showed higher percentages of linoleic acid than fruit pulp (41.2 and 29.2% respectively), but pulp presented higher percentages of oleic acid than peel (26.8% in pulp and 14.4% in peel). Both peel and pulp showed similar percentages of palmitic acid [37].

8. Health benefits: antioxidant activity

Antioxidant activity is one of the major mechanisms by which fruits and vegetables provide health benefits. Fruits and vegetable are also able to inhibit excessive oxidation due to free radicals, which are in the form of reactive oxygen species [9]. Prickly pear is rich in antioxidant product, containing phenolic compounds, carotenoids, betalains and vitamin C, all of which could be directly responsible for the health benefits [39]. Antioxidant activity in prickly pear fruit and peels may be affected by environmental factors, cultivar, genetic diversity, phenotype, agronomic practices, environmental and climatic conditions and processing of the fruit, among others [40]. Besides, the processing method and the extraction solvent affect antioxidant activity of *O. ficus-indica* extracts [26].

Antioxidant activity can be measured by different methods depending on the various mechanisms of antioxidant action. For example, some authors [2, 6, 8–10, 26, 41] studied antioxidant activity by DPPH, $ABTS^{·+}$, FRAP and ORAC methods. DPPH method consists in the elimination of DPPH radical by antioxidant

Method	Unit	Pulp	Peel	References
ABTS	mmol Trolox kg^{-1} dw	6.40–30.6	14.7–36.9	[9]
	μmol Trolox g^{-1} fw	6.70	—	[10]
DPPH	mmol Trolox kg^{-1} dw	58.4–60.1	54.8–59.6	[9]
	μmol Trolox 100 g^{-1} fw	108.85–122.47	141.60–141.80	[6]
FRAP	mmol Trolox kg^{-1} dw	15.0–32.3	40.2–116	[9]
	μmol Fe (II) g^{-1} dw	18.42–137.65	58.70–175.44	[26]
ORAC	mmol kg^{-1} fw	3.68–8.16	—	[41]
	μmol Trolox g^{-1} fw	26.3	—	[2]

dw, dry weight; fw, fresh weight.

Table 4.
Antioxidant activity of prickly pear fruit pulp and peel by different methods.

compounds present in the extracts, which determines its ability to capture radicals. The ABTS method captures the cationic $ABTS^{\cdot+}$ radical. FRAP method measures the ability to reduce Fe^{3+} in the sample. ORAC method measures the ability of the sample to scavenge peroxyl radicals.

Table 4 shows the antioxidant activity of *O. ficus-indica* depending on the method and the part analyzed (pulp and peel). The scavenging activity of DPPH, $ABTS^{\cdot+}$ and FRAP methods is higher in fruit peel. This trend can be observed in other fruits like pomegranate [42], guava fruit [43] and berries [44]. The consumption of fruits with high antioxidant activity, such as prickly pear fruits, is related to preventing degenerative diseases such as cancer, diabetes, hypercholesterolemia, arteriosclerosis or cardiovascular and gastric diseases [21, 25].

9. Processed products

One of the oldest ways to preserve highly perishable fruits is through different processing systems. Although it is necessary to do more research in preservation of prickly pear fruit and use it out of the harvest period, there are some processed products obtained from prickly pear fruit. The main ones are juices and nectars, marmalades and jams, dehydrated sheets, sweeteners, alcohol and wines [29, 45].

Juices and nectars from prickly pear fruit are mostly water. They contain appreciable amounts of sugars, vitamins and mineral salts (mainly potassium, calcium and sodium). They also are a good source of bioactive substances such as phenolic compounds, betalains, vitamin C and β-carotene. These products show different percentages of fruit pulp (15–75%), citric acid (0.3%), sucrose and water [45, 46].

Marmalades and jams are usually prepared from ripe fruits with high sugar content. In their manufacturing, it is important to control the sugar/pulp ratio, type and quantities of acidifying agents and the percentage of added pectin (thickening agent). Prickly pear fruit pulp already contains pectin, responsible for the viscosity of the pulp, which is a positive element toward the production of juices, marmalades and jams [45, 47].

Regarding prickly pear dehydrated sheets, there are different formulations and methods for their elaboration, mixing pulp in different sucrose ratios (0–10%). The thickness of the sheets is usually 5–15 mm. The preparations need to be spread and then dried at 60–70°C for at least 44 hours. Some authors mix prickly pear pulp with other fruits, like quince or melon pulps [45, 48, 49].

Sweetener liquid preparation from prickly pear fruit pulp implies enzymatic clarification of pulp juice, its decoloration and its vacuum concentration until 60°Brix (56% of glucose, 44% of fructose approximately). The obtained product shows a density and water activity similar to that of honey and marmalades and its characteristics are similar to other sweetener liquids currently marketed [45, 50].

Alcoholic beverages from *O. ficus-indica* are less known than those from other processed products. Some authors, for obtaining prickly pear wine, inoculated their juice with *Saccharomyces cerevisiae* and added SO_2 (10 mg L^{-1}) and citric acid for obtaining a pH 3.3, and then performed fractional distillation [51]. Besides, prickly pear fruit pulp can be added to other alcoholic beverages such as yakju, increasing the levels of alcohol, sugars and antioxidant activity [52].

Prickly pear seed oil is another potential product that can be obtained during fruit processing. Linoleic acid is the main fatty acid, and the percentages of PUFAs and MUFAs reach 63.29 and 23.30%, respectively [38]. Besides, other physical and chemical characteristics, such as refractive index, iodine number and saponification number, make it similar to other vegetable oils such as corn or grape seed oil [45].

1. Economic evaluation of prickly pear fruit production

Nowadays, *O. ficus-indica* cultivation is developed in at least 18 countries in arid and semiarid areas. The extension of this crop is more than 100,000 ha [53]. This does not include naturalized plants or plants cultivated for home consumption.

Prickly pear has been used since the sixteenth century as an important subsistence crop in many communities of Africa, Asia, Europe and America, although fruit consumption remains limited to local ethnic markets and there is little export. Only Mexico, Italy, Chile, South Africa and Argentina produce cactus pear in a commercial way [3].

Mexico is the world's largest producer of prickly pear, accounting for 45% of world production [3, 4]. Other important producing countries of prickly pear are Italy (12.2%) and South Africa (3.7%). The rest of the production is in Argentina, Chile, Bolivia, Peru, Colombia, United States of America, Morocco, Algeria, Libya, Tunisia, Egypt, Jordan, Pakistan, Israel, Greece, Spain and Portugal [3, 4].

Regarding Mexico, the planted area covers around 50,000–70,000 ha and the gross annual production is 300,000–500, 000 tones. It is the fifth fruit crop in the country and about 20,000 families obtain some income from cactus pear cultivation. Vegetable production, featured by small plots of land cultivation, supposes an additional 12,000 ha of cultivated area [54]. In this country, the cultivation of prickly pear presents the advantage that it produces employment and income in areas where few other crops can be produced [55].

Italy is the second world producer and the principal world exporter of cactus pear, mostly concentrated (96%) in Sicily with 7000–8300 ha producing about 78,000–87,000 tones per year [55]. South Africa's 1500 hectares produces about 15,000 tones. Other countries where cactus pear is cultivated are South Africa (1500 ha, 15,000 tones of fruit production), Argentina (1650 ha), Brazil (500,000 ha), Chile (934 ha), Peru (5000 tones of fruit production) and California (120 ha) [4]. However, it is difficult to quantify areas and production of prickly pear crop because it is a crop with low economic and social importance in most of the countries, so that there are not consistent economic data about it [4].

In Mexico, the main producer, the average production is approximately 12.8 t ha^{-1} (400 crates), which are sold at an average price of 3.2 euros each crate. This gives a total of 1280 euros per hectare, and the profit is approximately 340 per hectare, because the costs of tools, weeding, pruning, fertilization, fumigation, harvest and transport, among others [56]. In the case of Italy, the average production is approximately

15.1 t ha^{-1}, the incomes are 5.71 euros per hectare on average and the average profit per hectare is 1658.88 euros [55, 57]. In Spain, average production per hectare is 234 t ha^{-1}, and the average price is 1.42 euros per kilogram. Prices depend on the moment of the season and go from 1.8 euros per kilogram to 1.05 euros. This implies an average income of 555, 254.7 euros per hectare. So, average profit is 545, 801 euros per hectare.

Besides, prickly pear fruits show a high amount of compounds with biofunctional, nutraceutical and cosmetic properties, above crops like *Opuntia joconostle*, *Ziziphus jujube, Stenocereus pruinosus, Stenocereus stellatus* and *Punica granatum* [58–61]. However, no economic value analysis of the cactus pear cultivation based on obtention of these biofunctional, medicinal, nutraceutic and cosmetic compounds has been done. These compounds reach a value in the marketplace of 213.68 € per 20 mg in the case of kaempferol or 204.53 € per 10 mg in the case of isorhamnetin, and both are present in prickly pear fruit, among others.

10. Other aspects

Besides the health benefits of fruit consumption, *O. ficus-indica* presents other multiple applications in different areas:

- There are studies about the antigenotoxic capacity of the cladodes against mycotoxin zearalenone (mycotoxin F-2, produced by some species of *Fusarium*) in mice [62, 63].

- Cladodes of *O. ficus-indica* could be used to produce biofuels, specifically bioethanol and biogas [64, 65].

- Due to its clotting power, cladodes could be used as a natural coagulant to remove turbidity and color in raw waters, with a yield of 65 g of coagulant per kg of cladodes [66].

- Some studies showed that supplementing the feeding of goats with cladodes and fruit peels may be an important resource to reduce their water intake, without detrimental effects on digestion, growth and meat quality [67, 68].

- Pigments of red and purple prickly pear cultivars could be used in food industry as additives in products like sweets, desserts and dairy products. These additives were obtained by microencapsulation technique of betalains [69, 70].

- Due to its Crassulacean acid metabolism (CAM), *O. ficus-indica* has been studied for its ability of endure prolonged drought and CO_2 uptake, which can help to mitigate effects caused by desertification and global climatic change [71, 72].

- *O. ficus-indica* could be used in phytoremediation of contaminated soils with Se, Pb and other contaminant substances [73, 74].

Author details

Francisca Hernández García[1]*, Lucia Andreu Coll[1], Marina Cano-Lamadrid[2], David López Lluch[3], Ángel A. Carbonell Barrachina[2] and Pilar Legua Murcia[1]

1 Department of Plant Sciences and Microbiology, Research Group in Plant Production and Technology, Miguel Hernández University, Escuela Politécnica Superior de Orihuela, Alicante, Spain

2 Department of Agrofood Technology, Food Quality and Safety Research Group, Miguel Hernández University, Escuela Politécnica Superior de Orihuela, Alicante, Spain

3 Department of Agrienvironmental Economics, Miguel Hernández University, Escuela Politécnica Superior de Orihuela, Alicante, Spain

*Address all correspondence to: francisca.hernandez@umh.es

References

[1] Kiesling R. Origen, domesticación y distribución de Opuntia ficus-indica. Journal of the Professional Association for Cactus Development. 1998;**3**:50-59

[2] Kuti JO. Antioxidant compounds from four Opuntia cactus pear fruit varieties. Food Chemistry. 2004;**85**(4):527-533

[3] Reyes-Agüero JA, Aguirre JR, Carlín-Castelán F, González-Durán A. Diversity of wild and cultivated opuntia variants in the meridional highlands plateau of Mexico. Acta Horticulturae. 2013;**995**:69-74

[4] FAO, Food and Agricultural Organization. In: Inglese P, Mondragon C, Nefzaoui A, Saenz C, editors. Ecologia del cultivo, manejo y usos del nopal. Roma, Italy: FAO; 2018. pp. 1-174

[5] García-Cayuela T, Gómez-Maqueo A, Guajardo-Flores D, Welti-Chanes J, Cano MP. Characterization and quantification of individual betalain and phenolic compounds in Mexican and Spanish prickly pear (Opuntia ficus-indica L. mill) tissues: A comparative study. Journal of Food Composition and Analysis. 2019;**76**:1-13. DOI: 10.1016/j.jfca.2018.11.002

[6] Cano MP, Gómez-Maqueo A, García-Cayuela T, Welti-Chanes J. Characterization of carotenoid profile of Spanish Sanguinos and Verdal prickly pear (Opuntia ficus-indica, spp.) tissues. Food Chemistry. 2017;**237**:612-622. DOI: 10.1016/j.foodchem.2017.05.135

[7] Cherkaoui-Malki M, Nasser B, El Kebbaj M, Badreddine A, Latruffe N, El-Mostafa K, et al. Nopal Cactus (Opuntia ficus-indica) as a source of bioactive compounds for nutrition. Health and Disease. Molecules. 2014;**19**(9):14879-14901

[8] Feugang JM. Nutritional and medicinal use of Cactus pear (Opuntia spp.) cladodes and fruits. Frontiers in Bioscience. 2007;**11**(1):2574

[9] Andreu L, Nuncio-Jáuregui N, Carbonell-Barrachina ÁA, Legua P, Hernández F. Antioxidant properties and chemical characterization of Spanish *Opuntia ficus-indica* Mill . cladodes and fruits. Journal of the Science of Food and Agriculture. 2017;**98**:1566-1573

[10] Fernández-López JA, Almela L, Obón JM, Castellar R. Determination of antioxidant constituents in cactus pear fruits. Plant Foods for Human Nutrition. 2010;**65**(3):253-259

[11] Frati AC, Jiménez E, Ariza CR. Hypoglycemic effect of Opuntia ficus indica in non insulin-dependent diabetes mellitus patients. Phytotherapy Research. 1990;**4**(5):195-197

[12] Ennouri M, Fetoui H, Bourret E, Zeghal N, Guermazi F, Attia H. Evaluation of some biological parameters of Opuntia ficus indica. 2. Influence of seed supplemented diet on rats. Bioresource Technology. 2006;**97**(16):2136-2140

[13] Serra AT, Poejo J, Matias AA, Bronze MR, Duarte CMM. Evaluation of Opuntia spp. derived products as antiproliferative agents in human colon cancer cell line (HT29). Food Research International. 2013;**54**(1):892-901. DOI: 10.1016/j.foodres.2013.08.043

[14] El-Said NM, Ashraf IN, Sahar AR, Deraz SF. Prickly pear [*Opuntia ficus-indica* (L.) mill] peels: Chemical composition, nutritional value and protective effects on liver and kidney functions and cholesterol in rats. Functional Plant Science and Biotechnology. 2011;**5**:30-35

[15] El-Beltagi HS, Mohamed HI, Elmelegy AA, Eldesoky SE, Safwat G. Phytochemical screening, antimicrobial,

antiaxidant, anticancer activities and nutritional values of cactus (*Opuntia ficus indicia*) pulp and peel. Fresenius Environmental Bulletin. 2019;**28**(2):1545-1562

[16] Salim N, Abdelwaheb C, Rabah C, Ahcene B. Chemical composition of *Opuntia ficus-indica* (L.) fruit. African Journal of Biotechnology. 2009;**8**(8):1623-1624

[17] Garcia-Amezquita LE, Tejada-Ortigoza V, Heredia-Olea E, Serna-Saldívar SO, Welti-Chanes J. Differences in the dietary fiber content of fruits and their by-products quantified by conventional and integrated AOAC official methodologies. Journal of Food Composition and Analysis. 2018;**67**:77-85

[18] Mena P, Tassotti M, Andreu L, Nuncio-Jáuregui N, Legua P, Del Rio D, et al. Phytochemical characterization of different prickly pear (*Opuntia ficus-indica* (L.) Mill.) cultivars and botanical parts: UHPLC-ESI-MSn metabolomics profiles and their chemometric analysis. Food Research International. 2018;**108**:301-308

[19] Albano C, Negro C, Tommasi N, Gerardi C, Mita G, Miceli A, et al. Betalains, phenols and antioxidant capacity in cactus pear [*Opuntia ficus-indica* (L.) mill.] fruits from Apulia (South Italy) genotypes. Antioxidants. 2015;**4**(2):269-280

[20] Butera D, Tesoriere L, Di Gaudio F, Bongiorno A, Allegra M, Pintaudi AM, et al. Antioxidant activities of sicilian prickly pear (Opuntia ficus indica) fruit extracts and reducing properties of its betalains: Betanin and indicaxanthin. Journal of Agricultural and Food Chemistry. 2002;**50**(23):6895-6901

[21] Yeddes N, Chérif JK, Guyot S, Sotin H, Ayadi MT. Comparative study of antioxidant power, polyphenols, flavonoids and betacyanins of the peel and pulp of three Tunisian Opuntia forms. Antioxidants. 2013;**2**(2):37-51

[22] Mabrouki L, Zougari B, Bendhifi M, Borgi MA. Evaluation of antioxidant capacity, phenol and flavonoid contents of Opuntia streptacantha and Opuntia ficus indica fruits pulp. National Journal of Technology. 2015:2-8

[23] Scalbert A, Johnson IT, Saltmarsh M. Polyphenols: Antioxidants and beyond. The American Journal of Clinical Nutrition. 2005;**81**(1):215-217

[24] El-Mostafa K, El Kharrassi Y, Badreddine A, Andreoletti P, Vamecq J, El Kebbaj MS, et al. Nopal cactus (*Opuntia ficus-indica*) as a source of bioactive compounds for nutrition, health and disease. Molecules. 2014;**19**(9):14879-14901

[25] Galati EM, Mondello MR, Giuffrida D, Dugo G, Miceli N, Pergolizzi S, et al. Chemical characterization and biological effects of sicilian *Opuntia ficus indica* (L.) mill. Fruit juice: Antioxidant and antiulcerogenic activity. Journal of Agricultural and Food Chemistry. 2003;**51**(17):4903-4908

[26] Aruwa CE, Amoo S, Kudanga T. Phenolic compound profile and biological activities of southern African *Opuntia ficus-indica* fruit pulp and peels. Lebensmittel-Wissenschaft & Technologie. 2019;**111**(May):337-344. DOI: 10.1016/j.lwt.2019.05.028

[27] Stintzing FC, Schieber A, Carle R. Evaluation of colour properties and chemical quality parameters of cactus juices. European Food Research and Technology. 2003;**216**(4):303-311

[28] Graça Miguel M, Gago C, Valente R, Guerreiro A, Antunes D, Manhita A, et al. Qualitative evaluation of fruits from different *Opuntia ficus-indica* ecotypes/cultivars harvested in South Portugal. Journal of Food Biochemistry. 2018;**42**(6):1-11

[29] Corrales-garcía J, Andrade-rodríguez J, Bernabé-Cruz E. Response of six cultivars of tuna fruits to cold storage. Journal of the Professional Association for Cactus Development. 1997;**2**:160-168

[30] Vázquez-Araújo L, Burló F, Melgarejo P, Carbonell-Barrachina ÁA, Martínez JJ, Calín-Sánchez Á. Volatile composition and sensory quality of Spanish pomegranates (*Punica granatum* L.). Journal of the Science of Food and Agriculture. 2010;**91**(3):586-592

[31] Andreu-Coll L, Noguera-Artiaga L, Carbonell-Barrachina ÁA, Legua P, Hernández F. Volatile composition of prickly pear fruit pulp from six Spanish cultivars. Journal of Food Science. 2020;**85**:358-363

[32] Flath RA, Takahashi JM. Volatile constituents of prickly pear (*Opuntia ficus indica* Mill., de Castilla variety). Journal of Agricultural and Food Chemistry. 1978;**26**(4):835-837

[33] Arena E, Campisi S, Fallico B, Lanza MC, et al. Aroma value of volatile compounds of prickly pear. Italian Journal of Food Science. 2001;**13**(3):312-319

[34] Oumato J, Zrira S, Petretto GL, Saidi B, Salaris M, Pintore G. Volatile constituents and polyphenol composition of *Opuntia ficus - indica* (L.) mill from Morocco. Revue Marocaine des Sciences Agronomiques et Vétérinaires. 2016;**4**(3):5-11

[35] Rodríguez SA, Díaz S, Nazareno MA. Characterization of volatile organic compounds of Opuntia fruit pulp - changes in cactus fruit aroma as a consequence of pulp processing. Acta Horticulturae. 2015;**1067**:305-310

[36] Rodríguez-Cruz M, Tovar A, del Prado M, Torres N. Mecanismos moleculares de los ácidos grasos poliinsaturados y sus beneficios. Revista de investigacion clinica. 2005;**57**(3):457-472

[37] Andreu-Coll L, Cano-Lamadrid M, Sendra E, Carbonell-Barrachina Á, Legua P, Hernández F. Fatty acid profile of fruits (pulp and peel) and cladodes (young and old) of prickly pear [*Opuntia ficus-indica* (L.) Mill.] from six Spanish cultivars. Journal of Food Composition and Analysis. 2019;**84**:103294. DOI: 10.1016/j.jfca.2019.103294

[38] De Wit M, Hugo A, Shongwe N. Quality assesment of seed oil from selected cactus pear cultivars (*Opuntia ficus-indica* and *Opuntia robusta*). Journal of Food Processing & Preservation. 2017;**41**(3):e12898

[39] Jimenez-Aguilar D, Mujica-Paz H, Welti-Chanes J. Phytochemical characterization of prickly pear (Opuntia spp.) and of its nutritional and functional properties: A review. Current Nutrition & Food Science. 2014;**10**(1):57-69

[40] Moussa-Ayoub TE, Abd El-Hady ESA, Omran HT, El-Samahy SK, Kroh LW, Rohn S. Influence of cultivar and origin on the flavonol profile of fruits and cladodes from cactus *Opuntia ficus-indica*. Food Research International. 2014;**64**:864-872

[41] Stintzing FC, Herbach KM, Mosshammer MR, Carle R, Yi W, Sellappan S, et al. Color, betalain pattern, and antioxidant properties of cactus pear (Opuntia spp.) clones. Journal of Agricultural and Food Chemistry. 2005;**53**(2):442-451

[42] Calín-Sanchez Á, Figiel A, Szarycz M, Lech K, Nuncio-Jáuregui N, Carbonell-Barrachina ÁA. Drying kinetics and energy consumption in the dehydration of pomegranate (*Punica granatum* L.) arils and rind. Food and Bioprocess Technology. 2014;**7**(7): 2071-2083

[43] Marquina V, Araujo L, Ruíz J, Rodríguez-Malaver A, Vit P. Composición química y capacidad antioxidante en

fruta, pulpa y mermelada de guayaba (*Psidium guajava* L.). Archivos Latinoamericanos de Nutrición. 2008;**58**(1):98-102

[44] Oszmiański J, Wojdyło A, Lachowicz S. Effect of dried powder preparation process on polyphenolic content and antioxidant activity of blue honeysuckle berries (*Lonicera caerulea* L. var. kamtschatica). LWT- Food Science and Technology. 2016;**67**:214-222

[45] Saenz C. Processing technologies: An alternative for cactus pear (Opuntia spp.) fruits and cladodes. Journal of Arid Environments. 2000;**46**(3):209-225

[46] Lamia I, Zouhir C, Youcef A. Characterization and transformation of the *Opuntia ficus indica* fruits. Journal of Food Measurement and Characterization. 2018;**12**(4):2349-2357. DOI: 10.1007/s11694-018-9851-z

[47] Sawaya WN, Khatchadourian HA, Safi WM, AL-Muhammad HM. Chemical characterization of prickly pear pulp, *Opuntia ficus-indica*, and the manufacturing of prickly pear jam. International Journal of Food Science and Technology. 1983;**18**(2):183-193

[48] El-Samahy SK. El-Mansy HA. Bahlol HE. El-Desouky, AI. Ahmed AE. Thermal process time and sensory evaluation for canned cactus pear nectar. Journal of the Professional Association for Cactus Development. 2008;10(1)85-107

[49] Atef AM, Abou-Zaid N, Ibrahim I, Ramadan MT, Nadir A. Quality evaluation of sheets, jam and juice from prickly pear and melon blends. Life Scinces Journal. 2013;**10**(2): 200-208

[50] Sáenz C, Estévez AM, Sepúlveda E, Mecklenburg P. Cactus pear fruit: A new source for a natural sweetener. Plant Foods for Human Nutrition. 1998;**52**(2): 141-149

[51] Bustos OE. Alcoholic beverage from Chilean Opuntia Ficus. American Journal of Enology and Viticulture. 1981;**32**(3):228-229

[52] Cho I, Huh C, Kim Y. 신안산 손바닥선인장의 첨가비율 및 부위별에 따른 약주의 품질특성. Korean Journal of Food Preservation. 2010;**17**(1):36-41

[53] Inglese P, Basile F, Schirra M. Cactus pear fruit production. In: Nobel S, editor. Cacti: Biology and uses. Berkeley and Los Angeles, California: University of California Press; 2002. pp. 163-184

[54] Gallegos-Vazquez C, Mendez-GallegosS,Mondragon-JacoboC. Producción sustentable de la tuna en San Luis Potosí. 2013

[55] Timpanaro G, Foti VT. The structural characteristics, economic performance and prospects for the Italian cactus pear industry. Journal of the Professional Association for Cactus Development. 2014;**16**(January):32-50

[56] Losada HR, Vieyra JE, Luna L, Cortés J, Vargas JM. Economic Indicators, Capacity of the Ecosystem of Prickly Pear Cactus (*Opuntia megacantha*) and Environmental Services in Teotihuacan, México to Supply Urban Consumption. Journal of Agriculture and Environmental Sciences. 2017;**6**:85-91. DOI: 10.15640/jaes.v6n1a9

[57] Basile F, Foti VT, Timpanaro G. Comparative economic analyses between conventional and eco-compatible cactus pear cultivation in Italy. Acta Horticulturae. 2002;**581**:47-61. DOI: 10.17660/ActaHortic.2002.581.2

[58] Cortez-García RM, Ortiz-Moreno A, Zepeda-Vallejo LG, Necoechea-Mondragón H. Effects of cooking methods on phenolic compounds in Xoconostle (*Opuntia joconostle*). Plant Foods for Human Nutrition. 2015;**70**:85-90. DOI: 10.1007/s11130-014-0465-2

[59] Hernández F, Noguera-Artiaga L, Burló F, Wojdyło A, Carbonell-Barrachina ÁA, Legua P. Physico-chemical, nutritional, and volatile composition and sensory profile of Spanish jujube (*Ziziphus jujuba* Mill.) fruits. Journal of the Science of Food and Agriculture. 2016;**96**:2682-2691. DOI: 10.1002/jsfa.7386

[60] García-Cruz L, Dueñas M, Santos-Buelgas C, Valle-Guadarrama S, Salinas-Moreno Y. Betalains and phenolic compounds profiling and antioxidant capacity of pitaya (Stenocereus spp.) fruit from two species (*S. pruinosus* and *S. stellatus*). Food Chemistry. 2017;**234**:111-118. DOI: 10.1016/j.foodchem.2017.04.174

[61] Elfalleh W. Total phenolic contents and antioxidant activities of pomegranate peel, seed, leaf and flower. Journal of Medicinal Plants Resource. 2012;**6**(32):4724-4730. DOI: 10.5897/jmpr11.995

[62] Zorgui L, Ayed-Boussema I, Ayed Y, Bacha H, Hassen W. The antigenotoxic activities of cactus (*Opuntia ficus-indica*) cladodes against the mycotoxin zearalenone in Balb/c mice: Prevention of micronuclei, chromosome aberrations and DNA fragmentation. Food and Chemical Toxicology. 2009;**47**:662-667. DOI: 10.1016/j.fct.2008.12.031

[63] Zourgui L, Golli E El, Bouaziz C, Bacha H, Hassen W. Cactus (Opuntia ficus-indica) cladodes prevent oxidative damage induced by the mycotoxin zearalenone in Balb/C mice. Food and Chemical Toxicology 2008;46(5):1817-1824

[64] Sánchez GF. Pontencial del cultivo de la chumbera (*Opuntia ficus-indica* (L.) Miller) para la obtención de biocombustibles. Madrid, Spain: Doctoral dissertation, Universidad Politécnica de Madrid; 2012

[65] Pérez Cadena R, Arana Cuenca A, Alejandro Lizardi M, Alejandro Medina S, Espinosa T, Martínez A, et al. Selección de una variedad de cladodio de nopal (*Opuntia ficus-indica*) con potencial para la obtención de bioetanol. In: XVI Congreso Nacional de Biotecnología y Bioingeniería. Guadalajara, Jalisco, México; 2015

[66] Villabona A, Paz I, Martinez J. Characterization of *Opuntia ficus-indica* for using as a natural coagulant. Revista Colombiana de Biotecnología. 2013;**XV**(1):137-144

[67] Costa RG, Beltrão Filho EM. Ramos do Egypto Queiroga RDC, Suely Madruga M, Nunes de Medeiros a, de Oliveira CJB. Chemical composition of milk from goats fed with cactus pear (*Opuntia ficus-indica* L. miller) in substitution to corn meal. Small Ruminant Research. 2010;**94**(1-3):214-217. DOI: 10.1016/j.smallrumres.2010.08.001

[68] Vieira EL, Batista ÂMV, Guim A, Carvalho FF, Nascimento AC, Araújo RFS, et al. Effects of hay inclusion on intake, in vivo nutrient utilization and ruminal fermentation of goats fed spineless cactus (*Opuntia fícus-indica* Mill) based diets. Animal Feed Science and Technology. 2008;**141**(3-4):199-208

[69] Otálora MC, Carriazo JG, Iturriaga L, Nazareno MA, Osorio C. Microencapsulation of betalains obtained from cactus fruit (*Opuntia ficus-indica*) by spray drying using cactus cladode mucilage and maltodextrin as encapsulating agents. Food Chemistry. 2015;**187**:174-181

[70] Saénz C, Tapia S, Chávez J, Robert P. Microencapsulation by spray drying of bioactive compounds from cactus pear (*Opuntia ficus-indica*). Food Chemistry. 2009;**114**(2):616-622. DOI: 10.1016/j.foodchem.2008.09.095

[71] Nobel PS, Pimienta-Barrios E, Hernández JZ, Ramírez-Hernández BC.

Historical aspects and net CO_2 uptake for cultivated Crassulacean acid metabolism plants in Mexico. The Annals of Applied Biology. 2002;**140**(2):133-142

[72] Nobel PS, Valenzuela-Tapia A, Zañudo-Hernández J, Pimenta-Barrios E, Rosas-Espinoza VC. Young daughter cladodes affect CO2 uptake by mother cladodes of *Opuntia ficus-indica*. Annals of Botany. 2004;**95**(2):363-369

[73] Escobar-Alvarado LF, Vaca-Mier M, Rojas-Valencia MN. Hydrocarbon degradation and lead solubility in a soil polluted with lead and used motor oil treated by composting and phytoremediation. Bulletin of Environmental Contamination and Toxicology. 2018;**100**(2):280-285. DOI: 10.1007/s00128-017-2211-6

[74] Bañuelos GS, Lin ZQ. Cultivation of the Indian fig Opuntia in selenium-rich drainage sediments under field conditions. Soil Use and Management. 2010;**26**:167-175

Chapter 9

Novel Trend in the Use of *Opuntia* (Cactaceae) Fibers as Potential Feedstock for Material Science Applications

Faten Mannai, Ramzi Khiari and Younes Moussaoui

Abstract

Lignocellulosic fibers from *Opuntia* biomass, family Cactaceae, were mainly studied for their sustainability and cellulose content richness. This chapter highlights the current exploitation of *Opuntia* (Cactaceae) as potential feedstock for value-added applications such as reinforcement in composites and paper manufacturing. Cellulosic fibrous network fractions were isolated from different plant parts, and their fundamental properties, chemical and structural compositions, were analyzed, and the obtained results were discussed. The obtained fibrous networks were incorporated into two thermoplastic polymers; their enhancement properties and biodegradability have been studied. However, different recent methods of cellulose fiber extractions (pulping) and paper manufacturing have been investigated by testing two procedures of delignification: chemical and semi-chemical pulping process; these operations were followed by fibrous suspension characterizations and paper productions. The obtained results show the suitability of *Opuntia* (Cactaceae) for the new trend in ecological and green materials.

Keywords: *Opuntia*, fibrous networks, chemical composition, composites, pulping

1. Introduction

Opuntia (Cactaceae) is a cactus (non-forest and perennial plant) from tropical, subtropical, arid, and semiarid regions, which exists in the form of a shrub or a tree and has an original-look/unique morphology with a height of up to 5 m and produces a sturdy trunk as it ages [1, 2]. This particular species exhibits extraordinary water storage capacity and is known for their drought-tolerant characteristics (xerophytic) [1, 3]. A wide variety of this species and subspecies has been developed, distinguished by spiny or spineless cladodes, cladode shape, branching, fruit color, pulp color, epicuticle wax morphology, and many other properties [4–7]. Cactaceae is a great tree-like cactus formed by numerous up-flat branches (cladodes) [8, 9]. In branches, cellulosic fibrous tissues are slowly grown and arranged in parallel and fuse laterally with neighboring ones, forming a flat net-like structure [10], strongly similar to the cellular structure of *Luffa cylindrical* fibers [9]. This natural cellular structure is made up of an interconnected network of fibers struts, which form the edges (angle situated between two struts) and faces of cells, and possesses excellent mechanical behaviors in spite of its low density [11]. Their specific mechanical properties are due to the hierarchical composite organization [12, 13]. Cactaceae is mainly considered as a rich plant of natural food mineral, protein, vitamin, dietary fiber, and antioxidant compound which can represent an important product to prevent some health problems, such as diabetes, cancer, cataracts, macular degeneration, and neurological and cardiovascular diseases [7, 14–18]. The fruit syrup of *Opuntia* has a powerful antioxidant effect and exhibited effective antimicrobial activity against *Staphylococcus aureus* and *Staphylococcus epidermidis* [19]. Indeed, Cactaceae by product (cladodes, fruit peels, seeds, etc.) was used for non-food applications by testing their applicability to decontaminate wastewater through both the adsorption and coagulation–flocculation processes [8, 20]. It also will be a valuable resource for new applications and devoted to the future trends in terms of applications of natural fibers in different sectors. Furthermore, it is considered as one of the strongest and stiffest available lignocellulose fiber from renewable plant biomass [8, 20, 21]. Recently, the developing needs for new alternative green-based product have led to enlarge the discovery and the research of new renewable resources of natural fibers. This chapter addresses this research and gives an overview of potential exploitations of a new renewable non-woody lignocellulose source from plant biomass which is *Opuntia* (Cactaceae). It is interesting to point out there are only very few reports on the use of *Opuntia* fibers as raw material for paper manufacturing and as natural filler in reinforced polymer composite sectors, and they have been found to be the most interesting and discerning materials.

The pulp and paper industry, one of the largest and diversified industrial sectors in the world, produced every year more than 400 million tons of paper by different manufacturing methods using wood raw materials [22] and many types of non-wood raw materials such as bagasse (sugarcane fibers), cereal straw, bamboo, reeds, esparto grass, jute, flax, and sisal [23]. For this reason, the selection of suitable non-wood fibers is critical for the yield of fibrous fraction, ease of processing, quality, and cost of the final fiber-based product [24]. *Opuntia* (Cactaceae) was used for paper manufacturing as non-woody fibers by applying two different pulping processes. The first procedure is based on the utilization of semi-chemical treatment using a soft operation of chemical delignification in soda-hydrogen peroxide (soda-HP) mixture and mechanical grinding for fiber deliberation [25]. The second procedure is a chemical treatment with soda-anthraquinone mixture (soda-AQ) [26].

The development of fiber-reinforced composite material as an alternative over many conventional materials has been characterized by their eco-friendliness regarding the accumulation of plastic waste in the environment, durability, and its significant enhancement in the structural, mechanical, and tribological properties [27–30]. The natural fiber-reinforced polymer composites (FRPCs) were used to replace conventional metal/material and synthetic fiber/material in various applications in order to reduce weight and for energy conservation. Different kinds of natural fibers are mainly used for developing natural FRPCs with high specific properties, cost effectiveness, and renewability. Plant fibers include leaf fibers (sisal and abaca), bast fibers (flax, jute, hemp, ramie, and kenaf), grass and reed fibers (rice husk), core fibers (hemp, jute, and kenaf), seed fibers (cotton, kapok, and coir), and all other types, which may include wood and roots [31]. FRPCs are also classified according to their content, i.e., based polymeric material and the filler one. The based polymer, which binds or holds the filler material in structures, is termed as a matrix or a binder material, while the filler material is present in the form of sheets, fragments, particles, bundle, or whiskers of natural fiber [31]. Fibers can be placed unidirectionally or bidirectionally in the specific orientation into the

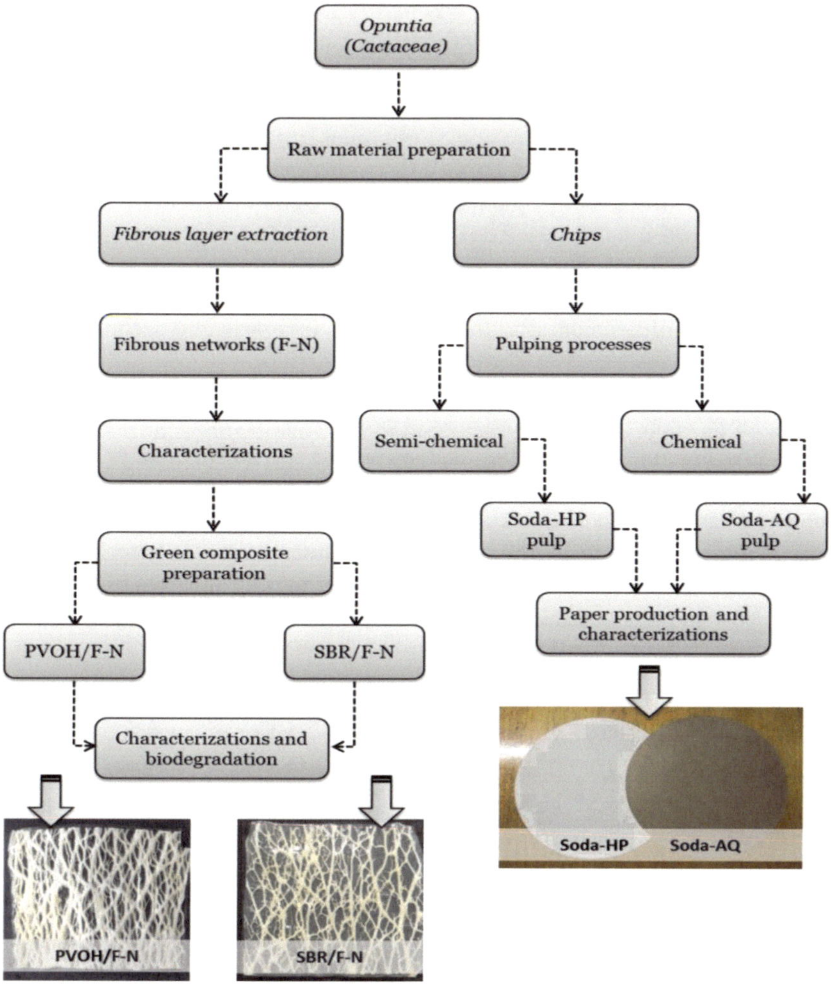

Figure 1.
The flowchart of (i) Opuntia *(Cactaceae) raw material obtained; (ii) pulping and paper manufacturing; and (iii) green composite elaborations.*

matrix structure, and they take loads from the matrix to the fiber in a very easy and effective way [31]. The arrangement and orientation of fibers define the properties and structural behavior of the composite material [32, 33]. *Opuntia* fibers were used as a natural filler to manufacture FRPC such as cactus fiber/polyester [21] and cactus fiber/polylactic acid [34]. This chapter provides an overview of the valorization and of *Opuntia* (Cactaceae) fibers in new green material science such as paper and bio-composite materials using two thermoplastic polymers which are polyvinyl alcohol and styrene butadiene rubber. The valorization ways of *Opuntia* (Cactaceae) fibers have been given in the flowchart in **Figure 1**.

2. Raw material characterizations: *Opuntia* (Cactaceae)

2.1 Fibrous layer extraction and characterizations: morphology, geometric dimensions, and mechanical behaviors

The isolations of fibrous network layers from *Opuntia* (Cactaceae) trunk using a green process in relation to their multifunctional features and its use as a raw material for novel ecological product was hardly studied for the first time by Mannai et al. [9]. **Figure 2** represents the fibrous networks (F-N) extraction steps which was performed manually and subsequently dried at room temperature for 7 days [9]. The obtained F-N layers (about 56 layers) represent a continuous phase (multidirectional fiber orientation angle) obtained from peripheral, middle, and central sections of the trunk, and **Table 1** displays their different characteristics.

Figure 3 shows the microscopic photograph of *Opuntia* fibrous layer obtained from the brightfield microscope. The obtained microscopic views show clearly the presence of axial primary fibers cross-linked by secondary ones. The bifurcation of primary fibers forming an open woven texture with special network design is also worth noting. The dimensions and forms of fibers (primary and secondary) have been related to the distribution of the layers in the trunk.

The F-N properties towards bulk density, morphological parameters including width, angles of opening pores, and area of pores of both fibers (primary and secondary) before and after swelling test, as well as the mechanical properties are listed in **Table 1**.

The peripheral section of the trunk regroups the thicker F-N layers than other sections of the *Opuntia* trunk. The average fiber width increases from the central to peripheral trunk sections and varies proportionally to the thickness of the fibrous layer. The average pore angle increases proportionally with the F-N pore area [9]. The pore angle between the primary fibers for the central section (90°) is 36%

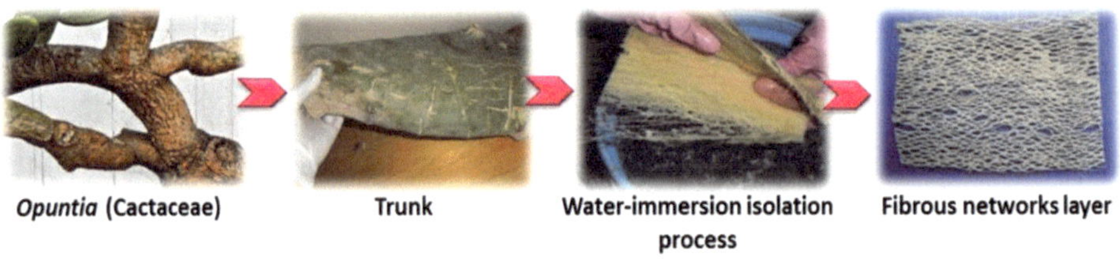

Figure 2.
Water-immersion process for fibrous networks layer extraction from the trunk of Opuntia *(Cactaceae).*

Layer sections	**Peripheral**				**Middle**				**Central**			
Apparent density (kg/m^3)	688–740				486–500				290–320			
Thickness (mm)	2.3–3.75				1.5–2.15				0.41–1.26			
Swelling ratio (%)	180 ± 12				135 ± 3				115 ± 5			
Geometric fiber dimensions	**Primary**		**Secondary**		**Primary**		**Secondary**		**Primary**		**Secondary**	
	Bs	**As**	**Bs**	**As**	**Bs**	**As**	**Bs**	**As**	**Bs**	**As**	**Bs**	**As**
Width (mm)	1.7	3.2	0.64	0.9	1.3	1.25	0.5	0.64	1	1.3	0.4	0.62
Pore angle (°)	54	42	25	20	80.3	68.3	41.1	37	90.7	59	41	26
Pore areas (mm^2)	2.8	1. 2	1.33	0.5	5.74	2.6	0.9	0.45	18.5	8	0.5	0.3
Mechanical structure	**Tensile**		**Flexural**		**Tensile**		**Flexural**		**Tensile**		**Flexural**	
Elastic modulus (GPa)	2.93		2.36		2.11		1.21		1.5		0.99	
Strength (MPa)	14.3		9.7		9.7		8.8		5.2		7.36	
Deformation at break (%)	5.04		6.18		1.7		4		1.4		2.9	

Table 1.
Apparent density, swelling ratio, and their effect on geometric dimensions before swelling (Bs) and after swelling (As) and mechanical strength properties obtained for fibrous networks obtained from peripheral, middle, and central sections of Opuntia *(Cactaceae) trunk.*

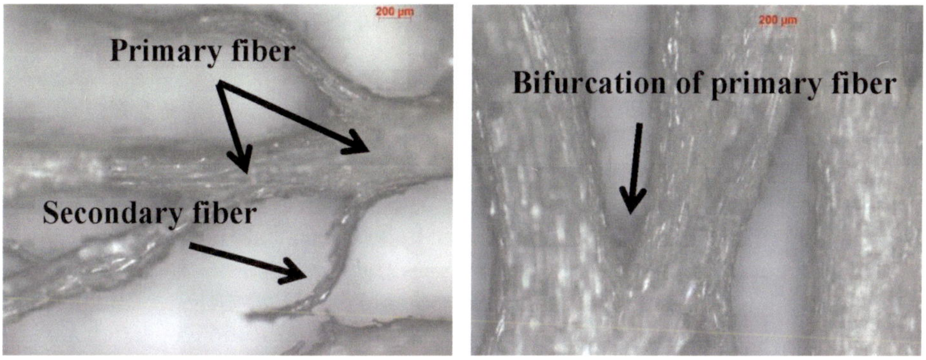

Figure 3.
Microscopic views of the surface of fibrous network layers from Opuntia *(Cactaceae) [200 μm].*

higher than the value obtained for primary fibers studied by Bouakba et al. [21] (about 57.5°); and for the secondary fibers, the pore angle of peripheral section is very acute (25°) compared to the other two sections (**Table 1**). The high obtained width with low pore area and low pore angle size of the outer layers of the trunk (peripheral section) confirm their dense structures which represent important fiber density with low porosity [9]. This finding is confirmed by the measured bulk density of these layers (see **Table 1**). The limited pore of the primary fibers of the central section (18.58 mm^2) was 69% and 84.5% higher than that of the middle and peripheral layers, respectively [9]. It was higher than that of the fibrous layers derived from the *O. ficus-indica* (11.3 mm^2) studied by Bouakba et al. [21].

Otherwise, the swelling ratio and uptakes by peripheral layers are higher than the middle and central ones; this could be due to the internal morphological aspect of *Opuntia* fibers which represent a porous structure, the presence of large fibro-vascular vessels [9, 25], and the high fiber density compared to those of the middle and central sections. The swelling of fibers can be explained by the hydrophilic nature of the Cactaceae plant, which can store a large amount of water [9]. Generally, the fiber hydration is noticed to be linked to the chemical composition of the fibers which have polar hydroxyl sites in their internal structures, which can form hydrogen bonds with water molecules [9]. The water-immersion process (applied for F-N extractions) could eliminate most of the water-soluble compounds (inorganic salts, ashes, coloring matter, etc.) from the fiber structure and favor the creation of void spaces, which could also explain the swelling of the fibers [9, 35].

The fiber water absorption can affect the geometric dimensions of *Opuntia* fibers by increasing the fiber width of both primary and secondary fibers (growth in size of the hydrated fibers) which can cause the decreasing of the pore areas and angles located between the primary and secondary fibers which may be explained by the occupation of the empty surfaces by the swelled fibers. Generally, highly hydrated fibers are characterized by their flexibility and ability to conform to fabric types [9].

The mechanical tensile and flexural behaviors of Cactaceae F-N were summarized in **Table 1**. It was found that the F-N tensile modulus increased from the central to the peripheral layers. The peripheral sections' tensile Young's modulus, uniaxial tensile strength, and deformation at breaks were found to increase compared to those obtained for the middle and central layers [9]. It is noted here that the peripheral layers have a favorably high Young's modulus compared to those of other cactus fibers [33–37].

The different flexural behaviors significantly increase from the central fiber layer to those of the peripheral F-N. This increase can be explained by the variation in geometric shape, layer thicknesses, fiber width, pore area distributions, fiber density, and bifurcation of primary fibers. It is worth noting that the flexural properties measured from the peripheral F-N layers are higher than those *O. ficus-indica* studied by Greco and Maffezzoli [34] and are lower than those found for *Myrtillocactus geometrizans* studied by Schwager et al. [36]. As expected, the F-N structural and geometric aspects modify the tensile and flexural states in such a way that the maximum elastic modulus shifts in an axial direction. This shift can be explained by the primary fiber orientation, which is axially aligned in most of the regions in the direction of the principal stresses and primary fiber density, on a macroscopic level. Mannai et al. [9] and El Oudiani et al. [38] affirmed and confirmed that the major factors that influence the F-N tenacity and elongation and give good mechanical properties include (i) the hierarchical structure; (ii) the unit cell dimensions (large and thick-wall parenchyma cells, long fiber bundles, and the densely distributed periderm with thick cell edges); and, on a microscopic level, (iii) the degree of crystallinity and (iv) the chemical composition of the fibers.

2.2 Chemical composition

Mannai et al. [25, 26] are the first ones to have studied the chemical composition of lignocellulosic fibers from *Opuntia* (Cactaceae) trunk. For comparison purposes, the results of this and other chemical compositions of *Opuntia* cladode studied by Malainine et al. [39] and some lignocellulosic raw materials from plant biomass collected from literature were summarized in **Table 2**.

A lower content of Klason lignin was observed in the *Opuntia* trunk and cladode and does not exceed 5 wt% (as opposed to other plants), indicating that *Opuntia* genus was a non-woody plant. The total holocellulose contents (64 wt%) were

Plant	Ash	K. lig	Holocel	α-cell
Opuntia (Cactaceae) trunk [25]	5.5	4.8	64.5	53.6
Opuntia (Cactaceae) cladode [39]	19.6	3.6	—	21.6
Date palm rachis [40]	5	27.2	74.8	45
Carrot leaves [41]	—	18.51	52.8	31.5
Rapeseed straw [42]	3.4	16	78.9	41.6
Amaranth [43]	12	13.2	58.4	32
Olive trimmings [44]	1	18.9	64.7	59
Softwood [45]	—	25–31	65–74	40–45
Harwood [45]	—	16–24	67–82	43–47
Alfa [46]	3.7	22.3	68.2	46.1
Eucalyptus citriodora [47]	0.8	22.7	—	48.2
Posidonia oceanica balls [40]	12	29.8	61.8	40
Vine stem [48]	3.9	28.1	65.4	35
Banana stem [49]	7.1	11.1	43.60	—
Annual plants [50, 51]	2–6.2	17–26	52–70	36–46

Table 2.
Chemical composition (ash; K. lig, Klason lignin; Holocel, holocellulose; and α-cell, α-cellulose) of Opuntia *(Cactaceae) trunk and other values obtained for cladode and their comparison with several lignocellulosic plants (w/w%).*

similar to that found in *olive trimmings*, hardwood, softwood, vine stems, and some annual plants; and it was clearly higher than those obtained for carrot leaves, amaranth, banana stems, and *Posidonia oceanica* balls; but it was lower than the holocellulose content measured for date palm rachis, rapeseed straw, and Alfa stems. In general, the holocellulose content can provide information about the quality and quantity of the produced pulp and paper [52]. The measured α-cellulose rate was surprisingly higher in the trunk (around 53.6 wt%) than those obtained for cladode (21.6 wt%) and other plants (**Table 2**); it was slightly lower than in *olive trimmings*. Non-wood fibers are handled in ways specific to their composition, and it was also acceptable for papermaking applications and corresponded to paper with enhanced strength [22]. For this reason, the processes used for the delignification of lignocellulosic fibers from *Opuntia* were adapted in very soft conditions to minimize degradation of the fibers and thus maximize pulp yield.

A very small fraction of inorganic compound (5.5 wt%) was observed in the trunk compared to the total mineral amount in the cladode, *Posidonia oceanica balls*, and banana stems; however, it was comparable to the values estimated for date palm rachis; but it was significantly higher than the ash contents measured for rapeseed straw, *olive trimmings*, Alfa stems, *Eucalyptus citriodora* and vine stems, and some annual plants (**Table 3**). The lower fraction of minerals in lignocellulosic fibers from the *Opuntia* trunk presents a major advantage, and the utilized raw material was silica free, which was extremely important for papermaking [25]. The chemical composition of ash was determined with elemental analysis and reported for the first time by Mannai et al. [25]. The resulting proportions, as seen in **Table 3**, are compared with other plants (amaranth, *Astragalus armatus*, date palm rachis, and banana pseudo stems) and have shown that the elemental composition of mineral contents in *Opuntia* can vary considerably from one species to another. A very low fraction of silicon (0.2 wt%) observed for *Opuntia* than those of other raw materials led to good separation after chemical delignification. It is clear that calcium and magnesium are the predominant inorganic materials in the Cactaceae family (18.33 and 16.54 wt%). The high presence of calcium due to the calcium oxalate crystals present naturally in *Opuntia* species [9, 25, 26]. The mineral elements present in this raw material do not present any counterindication for chemical pulping, composite manufacturing, and the area of the extraction of various cellulosic derivatives.

(%)	***Opuntia*** **(Cactaceae) [25]**	**Amaranth [43]**	***Astragalus armatus*** **[53]**	**Date palm rachis [40]**	**Banana pseudo stems [49]**
Si	0.2	0.25	18.42	2.8	2.7
Ca	18.33	4.17	11	21.5	7.5
Mg	16.54	0.035	2.90	3.53	4.3
Fe	399 ppm	—	0.29	240 ppm	—
Cu	192 ppm	0.01	<0.1	360 ppm	—
K	11.1	36.67	0.59	10.2	33.4
P	0.24	—	8.11	0.7	2.2
S	2.51	—	0.94	1.69	—
C	3.84	—	4.1	1.5	—
Na	0.6	—	1.8	6.79	—

Table 3.
Ash composition of Opuntia *(Cactaceae) trunk in comparison with data from previously published studies.*

3. Potential applications of cellulose fibers from *Opuntia* (Cactaceae)

3.1 Pulping and paper manufacturing

For papermaking, two main steps are followed in which the raw material is firstly cooked to obtain fibrous mass (pulp), and then the pulp is converted into paper. Mannai et al. [25, 26] were the first to find the preparation of pulp and papers from *Opuntia* trunk using semi-chemical and chemical pulping procedures, with yields of 80.8 and 41.1%, respectively [54]. Multistep pulping processes were followed to produce pulps and papers from *Opuntia* as shown in **Figure 1**. The manufacturing of pulp starts with raw material preparation [55], in which the dried *Opuntia* trunk was cut into chips (2–3 × 1–2 × 1.5–2 cm^3) [25, 26]. Two processes have already been applied to the delignification of *Opuntia* chips. The semi-chemical procedure based on the chemical treatment of raw material using soda–hydrogen peroxide (soda–HP) mixture (with the control of pH∼11) and the delignification reaction steps are done under reflux [25, 54]; these steps are followed by mechanical deliberation operation of cooked chips to more individualize and deliberate the fibrous suspensions. The obtained soda–HP pulp was purified by the classification of fibers by applying the standard T275 sp-12 method. Likewise, it has already been applied to the delignification of *Opuntia* trunk chips in a procedure described by Mannai et al. [26, 54], which utilized a total soda alkali charge of 20% (w/w o.d.) and an anthraquinone concentration of 0.1% (w/w o.d.). The liquor to solid ratio was kept at 10, and the mixture was cooked for 120 min at 170°C with a temperature ramping rate equal to 2.4°C/min. All of their experiments were conducted in a 1 L reactor that took 1 h to reach a constant temperature.

The morphological fiber's dimensions of the obtained fibrous suspensions in terms of their average length (mm) and width (μm) and the percentage of fine elements were examined using a MORFI (LB-01) analyzer developed by Techpap. The obtained results are summarized in **Table 4**. The fiber length (and width) of the *Opuntia* semi-chemical and chemical pulps were 764 μm (38 μm) and 737 μm (54.6 μm), respectively, which are in the same range of hardwood fibers [56]. The

Pulp and paper properties	Pulping process	
	Semi-chemical [25]	Chemical [26]
Yield (%)	80.8	41.4
Fiber length (μm)	764	737
Fiber width (μm)	38	45.6
Fine elements (%)	16.3	29.3
Bases weight (g/m^2)	38.4	65.2
Thickness (μm)	149	135
Bulk (cm^3/g)	2.26	2.07
Burst index (kPa m^2 g^{-1})	0.67	5.8
Tear index (mNm^2 g^{-1})	19.2	12
Young's modulus (GPa)	1.7	1.83
Breaking length (km)	1.9	1.57

Table 4.
Fiber and handmade paper produced from Opuntia *(Cactaceae) pulps after semi-chemical and chemical pulping procedures.*

processing with semi-chemical procedure gives a thick individualized fiber. It was considered as short fiber species [54]. It is also necessary to note that the pulp obtained from chemical procedures (at high temperature ~170°C) was characterized by a high content (29.3% of the length) of fine elements.

The semi-chemical and chemicals pulps obtained from *Opuntia* trunk after delignification were exploited to make hand sheets. Paper sheets have been successfully manufactured as shown in **Figure 1**. Papers from semi-chemical pulps are the whitish than the ones obtained from chemical pulp. This is explained by the treatment with hydrogen peroxide which oxidizes the color of chemical groups.

The given data of physical properties of hand sheet papers, as seen in **Table 4**, confirms that the studied raw material has potential for use in paper manufacturing using the soft delignification by applying the semi-chemical procedure which can affect the paper properties by increasing the fiber flexibility and strength [22, 25, 54]. Thus, these data suggest that *Opuntia* (Cactaceae) fibers can be used for producing paper from non-woody plants with various qualities (strength) and for future green product applications.

3.2 *Opuntia* (Cactaceae) fibrous network (F-N)-reinforced polymer composites: PVOH/F-N and SBR/F-N

The reinforcing potential of F-N obtained from *Opuntia* (Cactaceae) trunk was also investigated by Mannai et al. [57], and the flowchart in **Figure 1** shows the main manufacture steps. Composites filled with F-N from *Opuntia* (Cactaceae) seem to be promising materials for green applications. Natural plant fiber polymer composites are a composite material consisting of a polymer matrix embedded with natural fibers [58]. It is representing a promising domain of value-added products derived from low-cost and naturally occurring raw materials. The processing methods performed to synthesize bio-composites are mainly based on fiber type, form, and position. *Opuntia* F-N was used as a bidirectional filler with intricate structure; it is considered as a heterogeneous sheet filler. Two thermoplastic polymers, which were polyvinyl alcohol (PVOH) and styrene-butadiene rubber (SBR), were used as the matrix polymers. The hand lay-up molding processing of PVOH and SBR-based composites was chosen according to the networks form of fibrous layer of *Opuntia*. The reinforcing potential of fibrous networks in composites was investigated by evaluating their properties, and interfacial adhesions between polymer/fibers were studied. The major factor that affects the reinforcement composite properties is the bonding strength between fiber and polymer matrix in the composite.

The previous sections have provided some characters of *Opuntia* (F-N)-reinforced polymer composites obtained from dynamic mechanical analysis (DMA), thermogravimetric analyses (TGA), and biodegradation potential (BP). DMA was carried out by testing strips in axial (VF) and horizontal (HF) directions of incorporated fibers in order to understand the effect of additives and fillers on composites or filled materials [59]. As given in the previous work reported by Mannai et al. [57], the incorporation of fibers vertically for each matrix enhanced the storage modulus, especially for SBR-based composite. Otherwise, the relaxation process for composites reinforced with fibers oriented vertically is significantly higher than the one obtained for the filler oriented horizontally [57]; this can be explained by the elastic behavior of thicker axial fibers than other fibers interconnected with bifurcation ones [9] (see **Figure 3**). Mechanical interlocking and interfacial bonding adhesion are sensitive and can be improved by the natural fibers' surface roughness (**Figure 3**). The thermal behavior of *Opuntia* (F-N)-reinforced polymer composites was carried out using TGA in the conditions described in detail by Mannai et al. [57]. The main thermal data are summarized in

Composites	T (°C)	Discussion
PVOH/F-N [57]	30–120	Evaporation of residual moisture
	120–380	Degradation of hemicelluloses in lignocellulosic fibers and the elimination of hydroxide groups from PVOH in the form of water molecules
	380–520	Decomposition of chain segments of PVOH molecules and lignocellulosic compounds (cellulose and lignin)
SBR/F-N [57]	30–300	Evaporation of absorbed moisture and residual water in SBR latex
	300–570	Volatilization of SBR (styrene derivative) and lignocellulosic fibers (cellulose and lignin)

Table 5.
Thermal characteristics of PVOH and SBR-based composites reinforced with F-N layers of Opuntia *(Cactaceae) trunk obtained from TGA measurements.*

Table 5. From the discussion in **Table 5**, we can notice that F-N enhances the thermal properties of the used thermoplastic polymers.

The biodegradability potential (BP) (soil-burial test) of the matrix and produced composites were obtained by the mass retention technique, following the procedure outlined in literature [60, 61], and the results are given in the previous work [57]. The evolution of BP vs. time for the different materials after soil burial decreases gradually and tends to 93 and 86.6%, respectively, for PVOH/F-N and SBR/F-N [57]. These values are higher than those reported for PVOH/palm kernel shell powder bio-composites (20%) [62] and PVOH/corn starch films (40%) [63]. It should be mentioned that cellulosic fibers from *Opuntia* (Cactaceae) plant, PVOH, and SBR are biodegradable in nature, in which they may serve as a source of energy and carbon for specific microorganisms [64, 65]. From this study, eco-friendly *Opuntia*-derived fiber-reinforced polymer (thermoplastic) composites would be the materials for near future not only as a solution to the growing environmental threat but also as a solution to alleviating the uncertainty of the petroleum supply.

4. Conclusions

Opuntia (Cactaceae) is an alternative and sustainable plant for fiber production for semiarid and arid regions. It is in large quantities and considered as a useful source in most countries without forests. Indeed, natural fibers derived from Tunisian *Opuntia* (Cactaceae) plants prove their greatest potentials for use in paper manufacturing by applying two pulping processes which affects the pulp properties and paper characteristics and composite applications (as reinforcement in thermoplastic polymers) because of their excellent characteristics such as low density, high specific stiffness, good mechanical properties, biodegradability, eco-friendliness, and good thermal resistance. *Opuntia* fibers can be a good reinforcement candidate for high-performance biodegradable polymer composites. This study has paved the way of proposing this xerophyte genus as a suitable new resource of non-woody fibers at different fields and as an environmentally friendly alternative to green product.

Acknowledgements

The authors would like to express their deep gratitude to Mohamed Naceur Belgacem, Professor and Director of the Grenoble INP-Pagora and Agefpi, for his valuable advice and assistance, as well as to the Tunisian Ministry of Higher Education and Scientific Research for the financial support.

Conflict of interest

The authors declare no conflict of interest.

Author details

Faten Mannai[1], Ramzi Khiari[4,5,6] and Younes Moussaoui[2,3]*

1 Material Environment and Energy Laboratory (UR14ES26), Faculty of Sciences of Gafsa, University of Gafsa, Tunisia

2 Organic Chemistry Laboratory (LR17ES08), Faculty of Sciences of Sfax, University of Sfax, Tunisia

3 Faculty of Sciences of Gafsa, University of Gafsa, Tunisia

4 Faculty of Sciences, UR13 ES 63, Research Unity of Applied Chemistry and Environment, University of Monastir, Monastir, Tunisia

5 Department of Textile, Higher Institute of Technological Studies of Ksar Hellal, Tunisia

6 University of Grenoble Alpes CNRS, Grenoble, France

*Address all correspondence to: y.moussaoui2@gmx.fr

References

[1] Pérez-Molphe-Balch E, Santos-Díaz MDS, Ramírez-Malagón R, Ochoa-Alejo N. Tissue culture of ornamental cacti. Scientia Agricola. 2015;**72**:540-561. DOI: 10.1590/0103-9016-2015-0012

[2] Scalisi A, Morandi B, Inglese P, Lo Bianco R. Cladode growth dynamics in *Opuntia ficus-indica* under drought. Environmental and Experimental Botany. 2016;**122**:158-167. DOI: 10.1016/j.envexpbot.2015.10003

[3] Jimenez-Aguilar DM, Mujica-Paz H, Welti-Chanes J. Phytochemical characterization of prickly pear (*Opuntia spp.*) and of its nutritional and functional properties: A review. Current Nutrition and Food Science. 2014;**10**:57-69. DOI: 10.2174/157340131001140328120952

[4] Paiva PMG, De Souza IFAC, Costa MCVV, Santos ADFS, Coelho LCBB. *Opuntia sp.* cactus: Biological characteristics, cultivation and applications. Advances in Research. 2016; **7**:1-14. DOI: 10.9734/AIR/2016/26125

[5] Felker P, Rodriguez SDC, Casoliba RM, Filippini R, Medina D, Zapata R. Comparison of *Opuntia ficus-indica* varieties of Mexican and Argentine origin for fruit yield and quality in Argentina. Journal of Arid Environment. 2005;**60**:405-422. DOI: 10.1016/j.jaridenv.2004.06.003

[6] Medina EMD, Rodríguez EMR, Romero CD. Chemical characterization of *Opuntia dillenii* and *Opuntia ficus-indica* fruits. Food Chemistry. 2007;**103**:38-45. DOI: 10.1016/j.foodchem.2006.06.064

[7] Ayadi MA, Abdelmaksoud W, Ennouri M, Attia H. Cladodes from *Opuntia ficus-indica* as a source of dietary fiber: Effect on dough characteristics and cake making. Industrial Crops and Products. 2009;**30**:40-47. DOI: 0.1016/j.indcrop.2009.01. 003

[8] Nharingo T, Moyo M. Application of *Opuntia ficus-indica* in bioremediation of wastewaters. A critical review. Journal of Environment Management. 2016;**166**:55-72. DOI: 10.1016/j.jenvman.2015.10.005

[9] Mannai F, Elhleli H, Ammar A, Passas R, Elaloui E, Moussaoui Y. Green process for fibrous networks extraction from *Opuntia* (Cactaceae): Morphological design, thermal and mechanical studies. Industrial Crops and Products. 2018;**126**:347-356. DOI: 10.1016/j.indcrop.2018.10.033

[10] Mattheck C. Design in der Natur–Der Baum als Lehrmeister. Rombach Druck- und Verlagshaus (Hersteller) Auflage: 4., überarb. u. erw., Rombach Druck- und Verlagshaus (Hersteller); 2006. pp. 340

[11] Gibson LJ, Ashby MF. Cellular Solids: Structures and Properties. 2nd ed. Cambridge: Cambridge University Press; 1997

[12] Chen Q, Pugno NM. In-plane elastic buckling of hierarchical honeycomb materials. European Journal of Mechanics A/Solids. 2012;**34**:120-129. DOI: 10.1016/j.euromechsol.2011.12.003

[13] Fan H, Qu Z, Xia Z, Sun F. Designing and compression behaviors of ductile hierarchical pyramidal lattice composites. Material and Design. 2014; **58**:363-367. DOI: 10.1016/j.matdes.2014.01.011

[14] Kuti JO. Antioxidant compounds from four *Opuntia* cactus pear fruit varieties. Food Chemistry. 2004;**85**: 527-533. DOI: 10.1016/S0308-8146(03)00184-5

[15] Corral-Aguayo RD, Yahia EM, Carrillo-Lopez A, Gonzalez AG. Correlation between some nutritional components and the total antioxidant

capacity measured with six different assays in eight horticultural crops. Journal of Agricultural Food Chemistry. 2008;**56**:10498-10504. DOI: 10.1021/jf801983r

[16] Rodriguez-Garcia ME, de Lira C, Hernandez-Becerra E, Cornejo Villegas MA, Palacios-Fonseca AJ, Rojas-Molina I, et al. Physicochemical characterization of nopal pads (*Opuntia ficus-indica*) and dry vacuum nopal powders as a function of the maturation. Plant Foods and Human Nutrition. 2007;**62**:107-112. DOI: 10.1007/s11130-007-0049-5

[17] Hernández-Pérez T, Carrillo-López A, Guevara-Lara F, Cruz Hernández A, Paredes-López O. Biochemical and nutritional characterization of three prickly pear species with different ripening behavior. Plant Foods and Human Nutrition. 2005;**60**:195-200. DOI: 10.1007/s11130-005-8618-y

[18] Rodríguez-Félix A, Cantwell M. Developmental changes in composition and quality of prickly pear cactus cladodes (nopalitos). Plant Foods and Human Nutrition. 1988;**38**:83-93. DOI: 10.1007/BF01092314

[19] Dhaouadi K, Raboudi F, Funez-Gomez L, Pamies D, Estevan C, Hamdaoui M, et al. Polyphenolic extract of barbary-fig (*Opuntia ficus-indica*) syrup: RP-HPLC-ESI-MS analysis and determination of antioxidant, antimicrobial and cancer-cells cytotoxic potentials. Food Analytical Methods. 2013;**6**:45-53. DOI: 10.1007/s12161-012-9410-x 15

[20] Elhleli H, Mannai F, Elaloui E, Moussaoui Y. Nitrophenol removal from wastewater onto prepared activated carbon from *Opuntia ficus-indica*. In: Kallel A, Ksibi M, Ben Dhia H, Khelifi N, editors. Recent Advances in Environmental Science from the Euro-Mediterranean and Surrounding Regions (EMCEI 2017). Advances in Science, Technology and Innovation (IEREK Interdisciplinary Series for Sustainable Development). Cham: Springer; 2018. pp. 1245-4126

[21] Bouakba M, Bezazi A, Boba K, Scarpa F, Bellamy S. Cactus fiber/polyester biocomposites: Manufacturing, quasistatic mechanical and fatigue characterization. Composites Science and Technology. 2013;**74**:150-159. DOI: 10.1016/j.compscitech.2012.10.009

[22] Bajpai P. Chap. 1—General introduction. In: Pulp and Paper Industry: Microbiological Issue in Papermaking. 1st ed. Elsevier; 2015. pp. 1-10. DOI: 10.1016/B978-0-12-803409-5.00001-X

[23] Gullichsen J. Fiber line operations. In: Gullichsen J, Fogelholm CJ, editors. Chemical Pulping—Papermaking Science and Technology. Vol. 2. Finland: Fapet Oy; 2000. p. A19

[24] Anupam K, Lal PS, Bist V, Sharma AK, Swaroop V. Raw material selection for pulping and papermaking using TOPSIS multiple criteria decision making design. Environmental Progress and Sustainable Energy. 2014;**33**:1034-1041. DOI: 10.1002/ep.11851

[25] Mannai F, Ammar M, Yanez J, Elaloui E, Moussaoui Y. Cellulose fiber from Tunisian Barbary fig "*Opuntia ficus-indica*" for papermaking. Cellulose. 2016;**23**:2061-2072. DOI: 10.1007/s10570-016-0899-9

[26] Mannai F, Ammar M, Yanez JG, Elaloui E, Moussaoui Y. Alkaline delignification of cactus fibers for pulp and papermaking applications. Journal of Polymer and the Environment. 2018;**26**:798-806. DOI: 10.1007/s10924-017-0968-7

[27] Chukov D, Nematulloev S, Torokhov V, Stepashkin A, Sherif G,

Tcherdyntsev V. Effect of carbon fiber surface modification on their interfacial interaction with polysulfone. Results in Physics. 2019;**15**:1026344. DOI: 10.1016/j.rinp.2019.102634

[28] Lebreton LCM, van der Zwet J, Damsteeg JW, Slat B, Andrady A, Reisser J. River plastic emissions to the world's oceans. Nature Communications. 2017;**8**:15611. DOI: 10.1038/ncomms15611

[29] Linul E, Vălean C, Linul PA. Compressive behavior of aluminum microfibers reinforced semi-rigid polyurethane foams. Polymers. 2018;**10**:1298. DOI: 10.3390/polym10121298

[30] Yongxu D, Dong L, Libin L, Guangjie G. Recent achievements of self-healing graphene/polymer composites. Polymers. 2018;**10**:114. DOI: 10.3390/polym10020114

[31] Rajak DK, Pagar DD, Menezes PL, Linul E. Fiber-reinforced polymer composites: Manufacturing, properties, and applications. Polymers. 2019;**11**:1667. DOI: 10.3390/polym11101667

[32] Nair AB, Joseph R. 9-Eco-friendly bio-composites using natural rubber (NR) matrices and natural fiber reinforcements. In: Chemistry, Manufacture and Applications of Natural Rubber. Woodhead Publishing: Sawston; 2014. pp. 249-283. DOI: 10.1533/9780857096913.2.249

[33] Agarwal BD, Broutman LJ, Chandrashekhara K. Analysis and performance of fiber composites. In: Chemistry, Manufacture and Applications of Natural Rubber. 4th ed. Wiley; 2017. p. 576

[34] Greco A, Maffezzoli A. Rotational molding of biodegradable composites obtained with PAL reinforced by the wooden backbone of *Opuntia ficus-indica* cladodes. Journal of Applied Polymer Science. 2015;**132**:42447. DOI: 10.1002/app.42447

[35] Bismarck A, Askargorta AI, Springer J. Surface characterization of flax, hemp and cellulose fibers; surface properties and the water uptake behavior. Polymer Composites. 2002;**23**:872-894. DOI: 10.1002/pc.10485

[36] Schwager H, Haushahn T, Neinhuis C, Speck T, Masselter T. Principles of branching morphology and anatomy in arborescent monocotyledons and columnar cacti as concept generators for branched fiber-reinforced composites. Advanced Engineering Materials. 2010;**12**:B695-B698. DOI: 10.1002/adem.201080057

[37] Niklas KJ, Molina-Freaner F, Tinoco-Ojanguren C, Hogan CJ, Paolillo DJ. On the mechanical properties of the rare endemic cactus *Stenocereus eruca* and the related species *S. gummosus*. American Journal of Botany. 2003;**90**:663-674. DOI: 10.3732/ajb.90.5.663

[38] El Oudiani A, Chaabouni Y, Msahli S, Sakli F. Crystalline character of *Agave americana L*. fibers. Textile Research Journal. 2008;**78**:631-645. DOI: 10.1177/0040517508089757

[39] Malainine ME, Dufresne A, Dupeyre D, Mahrouz M, Vuong R, Vignon MR. Structure and morphology of cladodes and spines of *Opuntia ficus-indica*. Cellulose extraction and characterization. Carbohydrate Polymers. 2003;**51**:77-83. DOI: 10.1016/S0144-8617(02)00157-1

[40] Khiari R, Mhenni MF, Belgacem MN, Mauret E. Chemical composition and pulping of date palm rachis and *Posidonia oceanica*—A comparison with other wood and non-wood fibre sources. Bioresource Technology. 2010;**101**:775-780. DOI: 10.1016/j.biortech.2009.08.079

[41] Kopania E, Wietecha J, Ciechanska D. Studies on isolation of cellulose fibres from waste plante

biomasse. Fibres and Textiles in Eastern Europe. 2012;**96**:167-172

[42] Mousavi SMM, Hosseini SZ, Resalati H, Mahdavi S, Garmaroody ER. Papermaking potential of rapeseed straw, a new agricultural-based fiber source. Journal of Clean Products. 2013; **52**:420-424. DOI: 10.1016/j.jclepro.2013.02.016

[43] Fiserova M, Gigac J, Majtnerova A, Szeiffova G. Evaluation of annual plants (*Amaranthus caudatus L., Atriplex hortensis L., Helianthus tuberosus L.*) for pulp production. Cellular Chemistry Technology. 2006;**40**:405-412

[44] Jimenez L, Pérez I, Maestre F, Ferrer JL. Fabricacion de pastas celulosicas a partir de materias primas no madereras utilizando tecnologias menos contaminantes. InvTéc Papel. 1996;**129**:606

[45] Gharehkhani S, Sadeghinezhad E, Kazi SN, Yarmand H, Badarudin A, Safaei MR, et al. Basic effects of pulp refining on fiber properties—A review. Carbohydrate Polymer. 2015;**115**: 785-803. DOI: 10.1016/j.carbpol.2014.08.047

[46] Marrakchi Z, Khiari R, Oueslatic H, Mauret E, Mhenni F. Pulping and papermaking properties of Tunisian Alfa stems (*Stipa tenacissima*)—Effects of refining process. Industrial Crops and Products. 2011;**34**:1572-1582. DOI: 10.1016/j.indcrop.2011.05.022

[47] Anupam K, Swaroop V, Sharma AK, Lal PS, Bist V. Sustainable raw material selection for pulp and paper using SAW multiple criteria decision making design. Quarterly Journal of Indian Pulp and Paper Technical Association. 2015; **27**:67-76

[48] Mansouri S, Khiari R, Bendouissa N, Saadallah S, Mauret E, Mhenni F. Chemical composition and pulp characterization of Tunisian vine stems. Industrial Crops and Products. 2012;**36**: 22-27. DOI: 10.1016/j.indcrop.2011.07.036

[49] Sixta H. Handbook of Pulp. Weinheim: Wiley; 2008. pp. 2-19

[50] Ferhi F, Das S, Moussaoui Y, Elaloui E, Yanez JG. Paper from *Stipagrostis pungens*. Industrial Crops and Products. 2014;**59**:109-11413. DOI: 10.1016/j.indcrop.2014.05.015

[51] Ferhi F, Satyajit D, Elaloui E, Moussaoui Y, Yanez JG. Chemical characterization and suitability for papermaking applications studied on four species naturally growing in Tunisia. Industrial Crops and Products. 2014;**61**:180-185. DOI: 10.1016/j.indcrop.2014.07.001

[52] Sharma AK, Anupam K, Swaroop V, Lal PS, Bist V. Pilot scale soda-anthraquinone pulping of palm oil empty fruit bunches and elemental chlorine free bleaching of resulting pulp. Journal of Clean Products. 2015;**106**: 422-429. DOI: 10.1016/j.jclepro.2014.03.095

[53] Moussaoui Y, Ferhi F, Elaloui E, Ben Salem R, Belgacem MN. Utilization of *Astragalus armatus* roots in papermaking. BioResources. 2011;**6**: 4969-4978

[54] Mannai F, Elhleli H, Khiari R, Moussaoui Y. Cellulosic fibers from lignocellulosic biomass for papermaking applications. In: Food Preservation and Waste Exploitation. Croatia: IntechOpen; 2019. pp. 1-14. DOI: 10.5772/intechopen.88388

[55] Smook GA. Handbook for Pulp and Paper Technologists. Canada: Joint Textbook Committee of the Paper Industry of the United States and Canada. TAPPI; 1992. p. 425

[56] Jahan MS, Chowdhury DN, Islam MK. Pulping of dhaincha

(Sesbaniaaculeata). Cellulose Chemistry and Technology. 2007;**41**:413

[57] Mannai F, Elhleli H, Dufresne A, Elaloui E, Moussaoui Y. Opuntia (Cactaceae) fibrous network-reinforced composites: Thermal, viscoeelastic, interfacial adhesion and biodegradation behavior. Fibers and Polymers. 2020 in press

[58] Ku H, Wang H, Pattarachaiyakoop N, Trada M. A review on the tensile properties of natural fiber reinforced polymer composites. Composites Part B: Engineering. 2011;**42**:856-873. DOI: 10.1016/j.compositesb.2011.01.010

[59] Ammar M, Khiari R, Belgacem MN, Elaloui E. Biomatrix from *Stipa tenacissima* L. and its application in fiberboard using date palm rachis as filler. Journal of Renewable Materials. 2017;**5**:116-123. DOI: 10.7569/JRM.2016.634136

[60] Di Franco C, Cyras V, Busalmen J, Ruseckaite R, Vázquez A. Degradation of polycaprolactone/starch blends and composites with sisal fibre. Polymer Degradation and Stability. 2004;**86**: 95-103. DOI: 10.1016/j. polymdegradstab. 2004.02.009

[61] Kaith B, Jindal R, Jana A, Maiti M. Development of corn starch based green composites reinforced with *Saccharum spontaneum L.* fiber and graft copolymers—Evaluation of thermal, physico-chemical and mechanical properties. Bioresource Technology. 2010;**101**:6843-6851. DOI: 10.1016/j. biortech.2010.03.113

[62] Alias NF, Ismail H, Ab Wahab MK. Properties of polyvinyl alcohol/palm kernel shell powder biocomposites and their hybrid composites with halloysite nanotubes. BioResources. 2017;**12**: 9103-9117

[63] Bras J, Hassan ML, Bruzesse C, Hassan EA, El Wakil NA, Dufresne A. Mechanical, barrier, and biodegradability properties of bagasse cellulose whiskers reinforced natural rubber nanocomposites. Industrial Crops and Products. 2017;**32**:627-633. DOI: 10.1016/j.indcrop.2010.07.018

[64] Azahari NA, Othman N, Ismail H. Biodegradation studies of polyvinyl alcohol/corn starch blend films in solid and solution media. Journal of Physical Science. 2011;**22**:15-31

[65] Abraham E, Elbi PA, Deepa B, Jyotishkumar P, Pothen LA, Narine SS, et al. X-ray diffraction and biodegradation analysis of green composites of natural rubber/ nanocellulose. Polymer Degradation and Stability. 2012;**97**:2378-2387. DOI: 10.1016/j.polymdegradstab.2012.07.028